FERGAL

Data Communications for Engineers

Data Communications for Engineers

Michael Duck
Peter Bishop
Richard Read

University of North London

 Addison-Wesley Publishing Company

Harlow, England ● Reading, Massachussets ● Menlo Park, California ● New York
Don Mills, Ontario ● Amsterdam ● Bonn ● Sydney ● Singapore ● Tokyo ● Madrid
San Juan ● Milan ● Mexico City ● Seoul ● Taipei

Addison Wesley Longman Limited
Edinburgh Gate
Harlow
Essex
CM20 2JE
England

Cover designed by Designers and Partners, Oxford
and printed by The Riverside Printing Co. (Reading) Ltd.
Typeset by Logotechnics Sales & Marketing Ltd., Sheffield.

Printed in Great Britain by T. J. Press (Padstow) Ltd., Padstow, Cornwall.

First printed 1996.

ISBN: 0-201-42788-5

British Library Cataloguing-in-Publication Data
A catalogue record for this book is available from the British Library.

Library of Congress Cataloging-in-Publication Data applied for.

Preface

Aims and objectives

Data communications underpins all forms of distributed computer-based systems and may be defined as the process of sending data reliably between two or more communicating parties. Data itself may represent a wide range of information such as the contents of a computer file, a digitized image signal used for a videophone connection or telemetry measurements sent from a remote site, such as a water pumping station, to a central base for monitoring and control purposes.

Data communications has emerged from the fusion of the more traditional telecommunications industry and the newer, all pervasive, world of computing. Traditionally, telecommunications has concentrated its interests upon telephone-based communications. As computing has developed and the need to connect distant computer systems has also grown, so telecommunications has provided the necessary technology to support the needs of data transmission between computers. Equally, computers have needed to develop communications procedures and rules which are effective over the supporting telecommunications networks. Increasingly, the telecommunications industry has offered dedicated data communications services and networks to the computer industry. This has advanced to the point now that distinctions between telecommunications and computing are, in many areas, almost impossible to make. Indeed, modern telecommunications services themselves make great use of computers, for example, to control the switching of connections within telephone exchanges.

This book examines a wide range of techniques, technologies and systems used in the support of data communications. In particular, it addresses a variety of data transmission methods used to transfer data between physically distant locations. Several computer-based networks, which arrange for a number of users, perhaps many thousands, to be interconnected on either a permanent or temporary, switched, basis are dealt with. In order to support successful communication, a set of rules and procedures have been developed, many of which are embodied in internationally agreed standards. We shall be looking at the standards bodies and their standards and recommendations and, in later chapters, where they are used in practice.

The book has an emphasis on the engineering aspects of data communications rather than on specific issues relating to the uses to which such data communications networks are put. In general, applications and uses are predominantly software based, and frequently demand little or no understanding of the engineering aspects underpinning the networks. Indeed, as will be seen in Chapter 1, there is a very strong move towards breaking down the various functions or tasks of computer communications into a series of manageable subtasks, each of which may be defined as one of a series of layers. This inevitably means that some of the tasks, those closest to the user and the required applications, are distanced from data communications engineering.

Intended audience

This book has been written with two broad ranges of readership in mind:

(1) Students on Engineering or Computer Science courses who are following courses in Data Communications. The level of the text is suitable for both undergraduate degree students and final-year HND students.

(2) Professional engineers, managers and data communications system users. The book can be used as either an introductory text for personnel moving into the area of data communications or for updating the knowledge of engineers and managers who have previous experience in more traditional aspects of telecommunications.

Summary of chapters

Chapter 1 Introduction

This commences with a review of the development of data communications which, although of interest in its own right, is used as a vehicle to set the scene and introduce a number of basic concepts which are developed in later chapters. Having established how data communications has developed we then turn our attention to the different types of data and see some examples of where such data originates. The general concept of communications models is introduced as a tool to enable design, implementation and understanding of communications networks and systems. The case is then made for international standards and the major standards organizations, including the ISO, are discussed.

Chapter 2 Data transmission

This looks at the transmission of data over a communication channel and a number of basic concepts used later in the book are introduced. The different techniques used in the transmission of data are discussed as are the characteristics of the different media over which the data is transmitted. Finally, the range of common network configurations in which communications networks can be arranged is discussed with particular emphasis on their relative efficiencies.

Chapter 3 Information theory

Communications systems are concerned with the most efficient way of transferring information from a source to a destination. This chapter develops the theory of how the information content of a message can be measured and provides the rationale for the unit of information, the bit. The theory is then extended into establishing the average information content of a source known as entropy. The entropy is used in calculating the efficiency of a code which enables different source codes to be compared. The chapter goes on to investigate some of the most efficient source codes, known as optimum codes.

Chapter 4 Line codes

The primary function of line codes is to match the characteristics of the transmitted data to the media over which it is to be transmitted. The chapter describes the particular requirements for both metallic and optical systems. It then looks, in some detail, at the codes which are in current use. The importance of timing and synchronization in line codes is discussed and the chapter ends with a look at the use of scramblers to assist in such techniques.

Chapter 5 Modems

This chapter discusses the requirements for modems and describes their basic characteristics. The various modulation systems such as PSK, DPSK and QPSK are explained along with the reasons for their selection. The widely used V.24 interface is described and examples of different types of modem connections are given. A brief description of the more common modem standards is also provided.

Chapter 6 Error control

The basic impairments that can lead to errors in a transmission link are introduced. To overcome the effects of these impairments some form of error control is required. The basic principles of two types of error control, namely forward and feedback error control (also known as automatic repeat on request, ARQ) are discussed. The use of codes in forward error control is covered in some detail. Finally, the different feedback control strategies – stop-and-wait, go-back-n, and selective-repeat – are discussed.

Chapter 7 Link control and management

The concepts of link management and flow control are introduced. Link management involves the setting up and disconnection of a link. Flow control ensures that blocks of data, in the form of either frames or packets, are transmitted across a link in an orderly fashion. Link throughput and transmission efficiency are defined and the effect of ARQ on throughput is explored. A widely used protocol, the High-level Data Link Control (HDLC) protocol, is used to illustrate these concepts.

Chapter 8 Packet switched systems

This chapter looks at the ways in which switched data networks can be operated. The advantages and disadvantages of circuit switched and message switched systems are described. The main focus of the chapter is the X.25 packet switched network standard, with particular reference to its implementation in the United Kingdom. The chapter concludes with a brief look at the relatively new but increasingly popular frame relay systems.

Chapter 9 Local area networks

Local Area Networks (LANs) are computer networks, commonly comprising a number of PCs, confined to a relatively small, or local, physical area of no more than a few kilometres' radius. They are usually confined to a single site or complex. The chapter reviews the various techniques that enable stations to gain access to the physical medium used in the LAN. Some performance comparisons of these various techniques are made. The three IEEE 802 LAN standards are then discussed in detail. The chapter concludes with a discussion of two high-speed networks, the first being an optical fibre based LAN known as FDDI, and the second, the IEEE 802.6 Metropolitan Area Network (MAN) primarily used for interconnection of a number of LANs.

Chapter 10 Integrated services digital network

As the traditional Public Switched Telephone Network (PSTN) evolves to an all-digital network, a variety of nontelephony based services may now be provided by such an enhanced PSTN. This is known as an Integrated Services Digital Network (ISDN) which may be defined as digital, end-to-end connectivity between tele-phony-based customers. Such networks were conceived to provide high-quality telephony services using improved digital technology. In addition, ISDNs are able to offer what was originally regarded as high-speed data transmission on a circuit switched basis. The chapter commences with the development of the PSTN to an all-digital network. It continues by examining the technology of an ISDN and the new services it is able to offer in addition to traditional telephony services. Network operation and standards are considered and availability, both nationally and interna-tionally, is commented upon.

Chapter 11 Broadband networks

Broadband networks are a further development from ISDNs with the potential to support services requiring transmission speeds far in excess of those for telephony and many data applications. This chapter indicates the type of broadband services now appearing, especially advanced video and multimedia services, which demand transmission speeds in excess of current network capabilities. Asynchronous Transfer Mode (ATM) is introduced and it is shown how this offers a flexible and efficient network to support broadband services into the next century.

Chapter 12 Network management

The concepts introduced in Chapter 7 of link management and efficiency are extended, culminating in an introduction to integrated network management systems. Network management standards are discussed with particular emphasis on ISO and internet environments. Finally, a practical network management system is presented.

Appendices

Appendix A deals with the topic of queuing theory. A glossary of terms and a bibliography are also provided.

Acknowledgements

The authors wish to acknowledge the assistance of the following in the preparation of this book: Mark Akass, Director of Global Support, Scitor Ltd, for his helpful comments on Chapter 12 and also for permission to include material on the SITA-Vision network management product; Paul Kirkby, Technical Strategy Manager, Nortel Technology and Moira Stewart, Alcatel Network Systems, for reading some preliminary material and for their helpful and constructive suggestions for the improvement of the text; Professor Fred Halsall for his cooperation in permitting the use of some of his Glossary contained in *Data Communications, Computer Networks and Open Systems* (Halsall, 1996); and Pat Prigmore for his contribution towards wordprocessing much of the manuscript whilst in its early stages.

Michael Duck
Peter Bishop
Richard Read
London, March 1996

Contents

Introduction

The review of the development of data communications with which this chapter begins, although of interest in its own right, is a vehicle to set the scene and introduce a number of basic concepts which are developed in later chapters. We then turn our attention to the different forms of data and offer some examples of where such data originates. The general concept of communications models is introduced as a tool to enable design, implementation and understanding of communications networks and systems. Standards and the major standards organizations are then discussed. A communications model which is central to current network philosophy is the Open Systems Interconnection (OSI) reference model for which standards have both been produced and are under development. Later chapters explore how the model is implemented in practice, generally through a particular standard. The Institute of Electrical and Electronic Engineers, although not a standards organization in its own right, and its influence in the design of computer networks are discussed. The standards of various organizations are then compared.

1.1 The development of data communications

This brief overview of the development of data communications allows an appreciation of the more complex concepts of later chapters. The appearance of digital electronics in the early 1960s led to the widespread use of digital computers. Three main communities rapidly deployed computers: large financial institutions, such as banks and building societies, to automate and reduce staffing; universities, for the solution of complex problems; and large commercial organizations, to improve management and efficiency. Early computers, despite having relatively limited computing power compared with those of today, were expensive. Their use was therefore shared, often by many users. This raised the issue of how users gained **access** to computers. Computer access which is both secure and fairly apportioned continues to be a topic of great importance and will be developed further in later chapters.

A very early computer access arrangement involved the use of punched cards. Typically, a program was produced on some local facility and then stored on a paper card in binary form by punching holes to represent binary ones or zeros. The program itself was run on a computer some distance away, often many miles. The cards were therefore physically transported, often by road, to the computer which would then process the program using the cards. Any results were returned on cards in the same manner. This is an example of a **transport mechanism**, albeit non-electrical, which may be described as the mechanism to pass a message between two parties. Punched card operation is also an example of **off-line** operation, where there is no electrical connection between the equipment used to produce the programs and that used to process them. Compared with modern-day floppy disks, punched cards were physically extremely large and carried only a very small amount of data.

Another issue which appears in communication is that of a **protocol**. This is nothing more than a set of rules to govern an activity. Consider two people holding a conversation. A simple protocol often observed by some cultures is that only one person speaks at a time. A person wishing to speak waits for a pause in the conversation before commencing. Some computer networks also use this protocol which is called **stop-and-wait**. For a floppy disk, there clearly needs to be an agreed system for formatting and reading disks so that they can be interpreted and processed by a computer. In other words there needs to be a **communication protocol**.

Another key feature of a communication system is its **transmission rate** or **data rate**. Speed of transmission of data is measured in bits per second (bps) by dividing the total number of bits transmitted by the time taken to complete such a transfer. With punched-card systems the elapsed time between the point of production of programs and when they were processed by a computer could be of the order of hours or even days. Clearly, calculation of transmission speed in a system that used punched cards led to very, very slow transmission rates. A system that uses permanent electrical connections is **on-line** and can operate at transmission speeds of up to hundreds of millions of bits per second (Mbps).

The emergence of **modems** changed data communications dramatically. A modem converts digital data into analogue signals. This device allowed the digital signals of computers to be converted into analogue-type signals suitable for transmission over the then almost universally analogue circuits of the telephone network or **Public Switched Telephone Network** (PSTN). This allowed remote terminals to be connected electrically to a distant or **host** computer, either permanently or temporarily, by means of a dial-up telephone connection. This is an early example of on-line operation which paved the way for interactive operation in which the stations at either end of a link could interact. It would be an understatement to say this greatly speeded up data transmission and hence processing! On-line operation also opened up new opportunities and applications, such as the remote access of databases.

With the introduction of on-line operation, new protocols became necessary to format and control the flow of data. The possibility now existed of interconnecting a number different computers, each of which might operate using a different protocol. **Networking**, which is the technique of connecting two or more user systems together, was in its infancy, and the emergence of the need for standards soon became self-evident.

Computer terminal equipment is generally called **Data Terminal Equipment** (DTE). Modems may be used to interconnect a DTE to a telephone line. A modem is an example of equipment which terminates the end of a communication network. Such equipment is referred to as **Data Circuit-terminating Equipment** (DCE). The connections between a DCE and a DTE are an example of an **interface**. Standard interfaces have been defined which govern the physical and procedural arrangements for connection. We shall consider the details of some standard interfaces in Chapter 5.

By the mid-1960s networks had advanced to the point that instead of one DTE connecting with a single host computer, DTEs, or users, began communicating with each other. This enabled resource sharing and intercommunication to occur. To support such activity, users had to be assigned an **address** in the same way as a telephone line or house has a unique identity. Such addresses enabled a network to **route** messages successfully from a send station to receive station across a network. For communication networks to route connections they must contain switches to select the paths, or routes, that a connection must follow. Hence many networks are called **switched networks**.

Simple DTE to host computer communication commonly uses a continuous connection for the duration of the message transfer. This is known as **circuit switching**, where a physical or fixed path is dedicated to a pair of devices for the duration of the communication. A dial-up modem arrangement to connect a DTE to a host computer is a good example of a circuit-switched connection.

Experiments with new data networks began at the end of the 1960s and operated on a **packet** basis whereby data messages, unless very short, are subdivided into a number of separate blocks, known as packets, each containing hundreds of bits of data. Each packet may flow through the network independently. This means that each packet must contain some addressing information. Packets may be **connection-oriented**, such that a route through the network, known as a **logical connection**, is established (not necessarily a fixed end-to-end physical connection) for the duration of the message exchange and over which all subsequent packets flow. Connection-oriented operation is similar to circuit-switched operation. Having established a logical connection, data is then transferred, using as many packets as is necessary, until the data transfer is complete. The connection is then terminated.

Although connection-oriented operation remains very common, **connectionless** operation is now gaining in popularity. Here each packet is considered by the network as a unique entity, separate from any other packets, even though it may form part of the same message transfer. This means that each packet must contain full address information and cause switching to be set up and cleared on a per packet basis. It also means that a route must be established for every packet of a complete message. Although this is suitable for short messages consisting of only one or two packets, clearly for lengthy interchanges consisting of large numbers of packets, connection-oriented operation is more attractive. Packet switching will be explored in detail in Chapter 8. Public telephone and packet networks are generally operated on a national basis and cover a very large geographical area. It is for this reason that they are called **Wide Area Networks** (WANs).

The 1970s saw computers becoming both available and affordable to society as a whole and they began to appear in everyday industrial and commercial environments. Local intercommunication between computers may be provided by a WAN, be it a telephone or data network. However, local interconnection using a WAN is unwieldy and expensive. **Local Area Networks** (LANs) appeared in the mid-1970s. These networks often comprise a number of personal computers (PCs) confined to a relatively small, or local, physical area and are suitable for a single site or complex. LANs allow machines to be economically connected together in, say, an office, using a free-standing dedicated communication network. Where a connection is required between more distant users, for example users on separate LANs, connection may be made using an intermediate WAN.

The 1980s saw the development of **Metropolitan Area Networks** (MANs) which connect a number of sites and typically span a metropolitan area, or city. Clearly, much more traffic may occur than on a LAN, hence they require high-speed data links. Optical fibre, rather than a metallic conducting medium, is therefore generally used to support the high data rates of MANs. A MAN is often seen as a **backbone** network providing a high-speed data communication facility to which a number of LANs may be interconnected. Both LANs and MANs are dealt with in Chapter 9.

The 1990s have seen a rapid increase in popularity in **internetworking**. Just as packet switching grew out of a need to bring together a number of disparate systems into a network, there is now a need to bring together different networks. Although a user may not be aware of the existence of more than one network, the different networks that are interconnected in this way retain their own individuality and characteristics. Such a grouping of networks is often referred to as an **internet**.

1.2 Types and sources of data

In general, there are three main types of signal to be transmitted: speech, video and data. The last is almost exclusively computer information. Although speech and video signals are analogue in nature, the technology now exists to digitize virtually any signal at source. Once this process has occurred any signal, irrespective of the type of information it represents, may be regarded merely as data. Video signals are in the main found in broadcast TV, cable TV, satellite and surveillance systems. They are naturally analogue, although now capable of being digitized. Historically, and even today, many transmission systems continue to handle video signals in analogue form because the technology for all-digital processing of video has become available only within the last few years. Analogue operation continues in some instances because it provides adequate quality at an attractive price, for example in surveillance systems.

Although many different networks have evolved, there is increasing pressure, through the introduction of standards, for data transmission to operate at preferred speeds or rates. Speeds commonly used are 1200, 2400, 4800, 9600 and 19 200 bps or multiples, and submultiples, of 64 kbps. A higher range of speeds of 2, 8, 34, 140

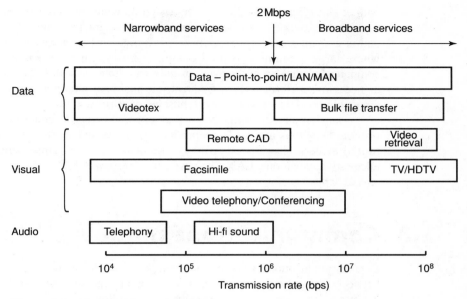

Figure 1.1 Types and sources of data.

or 565 Mbps is widely available in the long-haul communication links of WANs offered by **Post, Telephone and Telecommunications** (PTT) operators.

Figure 1.1 illustrates the major sources of data and their accompanying transmission rates. Additionally, the sources are categorized into audio, visual and data. Services operating at under 2 Mbps are **narrowband**, whereas higher rates are **broadband**. These topics will be developed further in Chapters 10 and 11, respectively.

Some of the sources are self-explanatory. Videotelephony has recently become commercially available in the form of a 'videophone' where an image of the telephony user appears at the distant videophone. Video conferencing is a method of holding a conference using TV cameras and monitors between two or more sites. It attempts to provide a conference facility and atmosphere without each participant having to travel to the same venue. Speech and images are relayed and close-ups may be included to view documents, photos and drawings.

High Definition Television (HDTV) is an extension of existing traditional TV services, but is of higher definition, leading to greater picture detail and sharpness. To facilitate improved information content a higher transmission rate is necessary. Videotex is a service which transmits alphanumeric and fairly simple graphics information. It is commonly found as an additional service offered within TV broadcast services or for remote booking services used, for instance, by travel agents for flight and holiday bookings. As indicated in Figure 1.1 Videotex has a relatively low transmission rate of up to 100 kbps or so.

Remote Computer Aided Design (CAD) allows CAD information to be transmitted. Such information occurs in a variety of CAD applications often associated with engineering design and may be design drawings or instructions for programs for example. Bulk file transfer transmission occurs at a fairly high transmission rate and is used to pass very large files of information between sites. An example is where large amounts of stored computer information are passed to a second computer for **backup** in the event of the first computer failing. These backups are, of necessity, performed quite frequently to maintain integrity, and because of the large volume of data involved they must be performed quite quickly.

Data has already been discussed in terms of the need to transmit at a distance, or more locally as in a LAN. Dependent upon the application, data may be transmitted at very slow speeds of, say, the order of tens of bps typical of some telemetry applications. LANs, however, have relatively high speeds of the order of tens of Mbps.

1.3 **Communications models**

Data communications is predominantly associated with supporting communications between two or more interconnected computer systems. Some of the main tasks necessary for successful communication are as follows:

- Initialization and release of a communications link.

- Synchronization between sending and receiving stations in order to interpret signals correctly, for example at the start and end of a packet.

- Information exchange protocols to govern communication. For instance a protocol needs to indicate whether a station may transmit and receive simultaneously or alternately.

- Error control to determine if received signals, or messages, are free from error. Where errors are evident some sort of corrective action should be instigated.

- Addressing and routing. These functions ensure that appropriate routing occurs within a network to connect a sending station to a receiving station with the correct address.

- Message formatting. This is concerned with the formatting or coding used to represent the information.

Although this list is not exhaustive, it does illustrate something of the range of complexity involved in a data transfer between communicating parties. In order to build a system, the communication task must be broken down into a number of manageable sub-tasks and their interrelationships clearly defined. A common approach to such analysis is to represent the tasks and their interrelationships in the form of a conceptual communications model. If the model is sufficiently well defined then the system may be developed successfully.

Although a given communications system may be modelled in a number of ways, computer communications models lean heavily towards **layered models**.

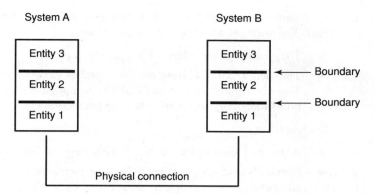

Figure 1.2   A layered communications model.

Each layer of the model represents a related series of tasks which is a subset of the total number of tasks involved in a communication system. Figure 1.2 illustrates a three-layer model connecting two systems.

A commonly used term in relation to communications models is that of an **entity**. An entity is anything capable of sending or receiving information. Examples of entities are user application programs, terminals and electronic mail facilities. Each layer of the model communicates using a layer-to-layer protocol between the two systems. This may be thought of as a communication between a pair of different programs, each operating within adjoining layers. Within a layered communication model, however, communication ultimately may only occur between the two systems via the physical connection. In order to support a protocol at a particular layer it must therefore operate via the lower layers of the model and then finally over the physical connection. The boundary between the lowest layer and the physical connection is a hardware interface. However, boundaries between the other layers are software in nature.

A well-constructed layered model enables complex systems to be specified, designed and implemented. This is achieved by the development of a model which splits these tasks into manageable layers and in such a way that the function of each layer is unique and can be implemented separately. Such a layered model then provides a framework for the development of protocols and standards to support information exchange between different systems. Although we have so far only considered layered models conceptually, in Section 1.6 we shall look in detail at the dominant layered model which is shaping data communications today.

1.4 Standards

Standards are required to govern the physical, electrical and procedural characteristics of communications equipment. Standards attempt to ensure that communications equipment made by different manufacturers will work satisfactorily with each other. The principal advantages of standards are:

- They ensure a large market for a particular product.
- They allow products from multiple vendors to communicate with each other. This gives purchasers more flexibility in equipment selection and use. It also limits monopoly and enables healthy competition.

Disadvantages are:

- A standard tends to lock or freeze technology at that point in time.
- Standards take several years to become established via numerous discussions and committee meetings, by which time more efficient techniques may have appeared.
- Multiple standards for the same thing often exist. This means that standards conversion is sometimes necessary. An example is the variety of TV standards throughout the world. In data communications an example is the USA's use of 1.5 Mbps digital cable transmission systems operated by PTTs compared with Europe's use of 2 Mbps systems.

The world of data communications is heavily regulated, legally and de facto. There exists a whole raft of standards bodies at international, regional and national level. Internationally, the two principal standards bodies concerned with data communications are the International Telecommunication Union (ITU) and the International Standards Organization (ISO).

1.4.1 International Telecommunication Union

The ITU is based in Geneva and is a specialized agency of the United Nations. It is comprised of member countries, each with equal status and voting rights, as well as other industrial companies and international organisations. On 1 July 1994 it was restructured. The new Telecommunications Standardization Sector (ITU-T) is responsible for setting standards for public voice and data services (formerly the remit of the Consultative Committee on International Telegraphy and Telephony, CCITT). The Radio Communications Sector (ITU-R) is responsible for Radio Frequency (RF) spectrum management for both space and terrestrial use. Both this and standards setting for radio used to be performed by the International Radio Consultative Committee (CCIR). The third sector of the ITU is the Development Sector (ITU-D) which is responsible for improving telecommunications equipment and systems in developing countries.

Each sector also organizes conferences on a world and/or regional basis and operates study groups. Standards are produced as a result of such activity to govern interworking and data transfer between equipment at an international level, rather than within the confines of a single nation. ITU-T recommendations which have been produced for use in data communications are:

- V-series – data transmission over telephone circuits (modems)
- X-series – packet and switched data and more recently all seven layers of the OSI reference model
- I-series – Integrated Services Digital Network (ISDN) transmission
- G-series – higher order multiplexed digital transmission systems
- Q-series – ISDN switching and signalling systems

Although, strictly speaking, these are recommendations, they are to all intents and purposes standards.

Later chapters will explore many of the above standards and make reference to the relevant standard numbers as appropriate.

1.4.2 International Standards Organization

The ISO promotes the development of standards in the world to facilitate the international exchange of goods and services. Its sphere of interest is not merely confined to data communications; for instance, the ISO produces specifications for photographic film. The organization is made up of members from most countries, each representing the standards bodies of their parent country, for example BSI for the United Kingdom and ANSI for the United States. The main involvement of the ISO in data communications has been its development of the reference model for open systems interconnection (OSI), which is discussed in Section 1.6. In the area of data communications, the ISO standards are developed in cooperation with another body, the International Electrotechnical Committee (IEC). Since the IEC is primarily interested in standards in electrical and electronic engineering, it tends to concentrate on hardware issues, whereas the ISO is more concerned with software issues. In the area of data communications (and information technology), in which their interests overlap, they have formed a Joint Technical Committee which is the prime mover in developing standards.

1.4.3 Other standards bodies

Regionally within Europe, the European Conference of Posts & Telegraphs (CEPT) has, until recently, been active in setting standards for use within Europe and has contributed towards the standards-making activities of the ITU-T. The European Telecommunications Standards Institute (ETSI), founded by the European Union, was formed in 1988 and is based in Nice. It comprises over 160 members from more than 20 countries throughout Europe within the EU and European Free Trade Association (EFTA) areas. ETSI has taken over the development of standards from CEPT for regulatory purposes in Europe. European Telecommunication Standards govern telephone services, packet switched data networks, Videotex and digital cellular services. CEPT still exists as a forum for strategic planning.

1.5 Open systems interconnection

As computer networks have proliferated, so the need to communicate between users located on different networks has emerged. Such intercommunicating computer systems are termed **distributed** computer systems and are required to process information and pass information between them.

Historically, communication between groups of computers and DTEs was generally restricted to equipment from a single manufacturer. Many systems use either IBM's Systems Network Architecture (SNA) or DEC's Digital Network Architecture (DNA) which are not directly compatible with each other. ISO formulated its open systems interconnection reference model in the late 1970s specifically to address the problem of **interconnectivity** between different user systems.

Open systems gives users of data networks the freedom and flexibility to choose equipment, software and systems from any vendor. It aims to sweep away proprietary systems which oblige a user to build a system with kit from a single vendor. Open systems is a concept which relies upon the emergence of common standards to which components and systems must conform. In this way full inter-connectivity occurs between users who may purchase equipment supplied by vendors of their choice.

1.5.1 *N-layer service*

The OSI reference model may be thought of as a series of conceptual layers. Such layers operate on the principle shown in Figure 1.3. Layer N provides service N to layer $N+1$. In order for layer N to fulfil its service to layer $N+1$, it requests a service from layer $N-1$. The interfaces or boundaries between adjacent layers or services are known as N-layer **Service Access Points** (NSAPs).

1.5.2 Peer-to-peer protocol

The concept of an N-layer service is intended to break down the complex tasks of networking into a series of logical and ordered subtasks, each of which becomes relatively simple to design and implement. Another plank in this process of decomposition of the networking task is the concept of **peer-to-peer** communication protocols whereby any given layer 'talks' to its corresponding layer at the distant end.

1.5.3 Encapsulation

We have established that protocols operate horizontally between layers in peer-to-peer fashion. However, to support such protocols, communication must in reality

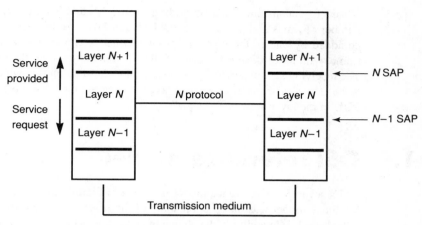

Figure 1.3 *N*-layer service.

occur up and down through the layers. The question then arises: how are protocols implemented between layers and also across the network? The answer is illustrated in the seven layer model of Figure 1.4. When an application has a message to send, data is sent to the top layer which appends a header (H). The purpose of the header is to include the additional information required for peer-to-peer communication. The resultant header and data are termed a **Protocol Data Unit** (PDU). Each PDU

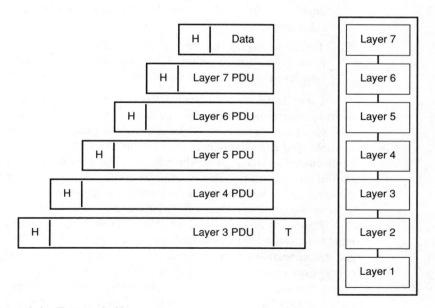

Figure 1.4 Encapsulation.

is said to encapsulate data by adding such a header. This process is repeated a number of times and generates a PDU at each layer. Additionally a trailer (T) may be added by one of the layers to produce what is known as a **frame**. Finally, the frame, with all of the encapsulated PDUs, is transmitted over the physical medium. The receiving station performs the reverse of encapsulation (known as decapsulation) as headers are stripped off at each layer to separate the respective header from the PDU of the layer above.

1.6 OSI reference model

OSI's seven-layer reference model is shown in Figure 1.5. The development of the reference model was based upon some of the principles discussed in general terms in Section 1.3. In addition, the choice of seven layers has sprung from the following:

(1) Only sufficient layers have been agreed such that each layer represents a different and unique function within the model.

(2) A layer may be thought of as the position of a related group of functions within the model which ranges from one specific to the user application at one end to one involving the transmission of data bits at the other. Layers between these two extremes offer functions which include interaction with an intermediate network to establish a connection, error control and data formatting.

(3) A layer should be so organized that it may be modified at a later date to enable new functions to be added and yet not require any changes within any other layer.

(4) The seven layers and their boundaries have attempted to build upon other models which have proved successful and in such a way as to optimize the transfer of information between layers.

Layer 1, the physical layer defines the electrical, mechanical and functional interface between a DCE and the transmission medium to enable bits to be transmitted successfully. The layer is always implemented in hardware. A common example used extensively in modems is the ITU-T V.24 serial interface which will be discussed in detail in Chapter 5. No error control exists at layer 1 but **line coding** may be incorporated in order to match data signals to certain properties of the communication channel. An example might be to remove a d.c. component from the signal. Line coding will be explored in detail in Chapter 4.

Layer 2 is the data link layer, the function of which is to perform error-free, reliable transmission of data. Link management procedures allow for the setting up and disconnection of links as required for communication. Having established a connection, error detection, and optionally error correction, is implemented to ensure that the data transfer is reliable. Flow control is also performed to provide for the orderly flow of data (normally in the form of packets) and to ensure that it is not lost or duplicated during transmission.

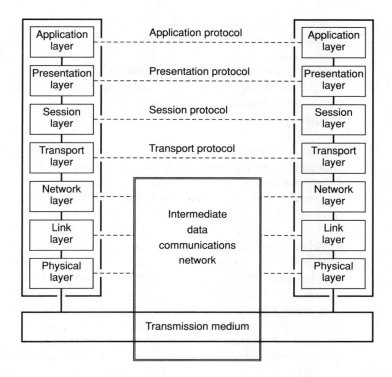

Figure 1.5   OSI reference model.

Layer 3 is the network layer, whose principal task is to establish, maintain and terminate connections to support the transfer of information between end systems via one or more intermediate communication networks. It is the only layer concerned with routing, offering addressing schemes which allow users to refer unambiguously to each other. Apart from the control of connections and routing, the layer, by engaging in a dialogue with the network, offers other services, such as a user requesting a certain quality of service or reset and synchronization procedures.

The intermediate data communications network included in Figure 1.5 is not strictly a part of the reference model. It is included to draw attention to the fact that the network may be regarded as a black box. What happens within it does not necessarily have to conform to any standards, although in practice this is rarely the case. Rather, what is important is that interfaces and protocols be fully supported at the network boundaries to ensure compatibility and interworking.

All of the lower three layers are heavily network-dependent, for example ITU-T's X.25 recommendation for gaining access to packet-switching networks specifies operation at layers 1, 2 and 3 only.

Layer 4 is the transport layer and separates the function of the higher layers 5, 6 and 7 from the lower layers already discussed. It hides the complexities of data communications from the higher layers which are predominantly concerned with supporting applications. The layer provides a reliable end-to-end service for the transfer of messages irrespective of the underlying network. To fulfil this role, the transport layer selects a suitable communications network which provides the required quality of service. Some of the factors which the layer considers in such selection are throughput, error rate and delay. Furthermore, the layer is responsible for dividing messages into a series of packets of suitable size for onward transmission through the selected communications network.

Layer 5, the session layer, is responsible for establishing and maintaining a logical connection. This may include access controls such as log-on and password protection. Secondly, the session layer performs **dialogue management**. This is merely a protocol used to order communication between each party during a session. For example, consider an enquiry/response application such as is used for airline ticket booking systems. Although two-way communication is necessary for such an interactive application it need not be simultaneous. Suppose that the connection provides communication in only one direction at a time. The protocol must therefore regulate the direction of communication at any one instant. If, however, full simultaneous two-way communication is available then little dialogue management is required save some negotiation at set-up time. The third, and most important, function of the session layer is recovery (or synchronization). Synchronizing points are marked periodically throughout the period of dialogue. In the event of a failure, dialogue can return to a synchronizing point, restart and continue from that point (using backup facilities) as though no failure had occurred.

Layer 6 is the presentation layer and presents data to the application layer in a form that it is able to understand by performing any necessary code and/or data format conversion. Therefore, the application layer need not be aware of the code used in the peer-to-peer communication at the presentation layer. It means that, in practice, users may operate with entirely different codes at each end which may in turn be different again from the code used across the network for intercommunication. **Encryption** may also be added at layer 6 for security of messages. Encryption converts the original data into a form which is ideally unintelligible to any unauthorized third party. Such messages may usually be decrypted only with knowledge of a key which must be kept secure.

Layer 7, the application layer, gives the user access to the OSI environment. This means that the layer provides the necessary software to offer the user's application programs a set of network services, for example an e-mail service. It is effectively the junction between the user's operating system and the OSI network software. In addition, layer 7 may include network management, diagnostics and statistics gathering and other monitoring facilities.

Most standards activity has centred on the lower layers to support communication networks and their interfaces, for example ITU-T's X.25 recommendation for packet-switched network operation addresses layers 1, 2 and 3, only. ISO standards have more recently addressed this imbalance with standards for some applications now available at all seven layers to support a truly open systems interconnection.

1.7 Institute of Electrical and Electronics Engineers 802 standards

The Institute of Electrical and Electronics Engineers (IEEE) is a US professional society and is not a standards body in its own right. Its major influence on international standards is through its IEEE 802 project which produced recommendations in 1985, initially for use with LANs, which were adopted as standards by the ISO in 1987. Since then further standards have been developed for MANs. Work continues in the development of new standards via Technical Advisory Groups. The current activities of the IEEE 802 committee are as follows:

- 802.1 Overview of 802 standards
- 802.2 Logical Link Control
- 802.3 CSMA/CD
- 802.4 Token Bus
- 802.5 Token Ring
- 802.6 MANs
- 802.7 Broadband LANs
- 802.8 Fibre Optic
- 802.9 LAN Integration
- 802.10 LAN Security
- 802.11 Wireless LANs

Logical link control (802.2) specifies the link control procedures for the correct flow and interchange of frames for use with the three basic LAN types. These are distinguished by the way in which a user gains access to the physical medium of the LAN, a process known as **Medium Access Control** (MAC), and are known as CSMA/CD, token bus and token ring (802.3, 4 and 5, respectively). The differences between these types of LAN relate fundamentally to network topology and access methods, signalling technique, transmission speed and media length. In Chapter 9 we shall examine medium access control in some depth.

1.8 OSI reference model and other standards

Table 1.1 makes comparison between the OSI reference model and various other standards for a variety of applications. In many instances the other standards are comparable with and sometimes identical to those of the ISO.

Table 1.1 Comparison of OSI layers with other standards.

OSI layers	OSI	ITU-T	DoD	IEEE
7. Application	CCR, FTAM, JTM,VT	X.500 (Directory Services) X.400 (Message Handling Services) ACSE, ROSE, RTSE		
6. Presentation	Presentation			
5. Session	Session			
4. Transport	Transport		TCP	
3. Network	IP	I-series (ISDN)	IP	
2. Data link	8802.2	X.25 (Packet Switching) X.21 (Circuit Switching)		802.2 Logical Link Control
1. Physical	8802.3/4.5			802.3/4/5

Some texts also include DNA and SNA in comparisons with ISO but as they are not standards they are omitted here. ISO itself has the following standards to support applications at the application layer:

- FTAM – File Transfer and Access Management
- VTs – Virtual Terminals
- CCR – Commitment, Concurrency and Recovery
- JTM – Job Transfer, Access and Management

ISO also has standards to implement the remaining protocol layer functions. The LAN standards 802.2 – 802.5 have been adopted by ISO as 8802.2–8802.5, respectively. The data link layer standard is 8802.2. Similarly, at the physical layer, MAC procedures and physical interconnection are offered as ISO 8802.3–5.

ITU-T supports a number of application-layer services including:

- X.500 – Directory Services
- X.400 – Message Handling Services
- RTSE – Reliable Transfer
- ROSE – Remote Operations
- ACSE – Association Control

ITU-T also has standards defining operation at the lower three layers for ISDN, packet switching and circuit switching.

The OSI reference model seeks to set out a method of developing inter-communication or internetworking via one or more, often different, computer communication networks. Additionally, having established the reference model, ISO is committed to developing a full range of standards to enable such inter-networking to be established effectively and easily.

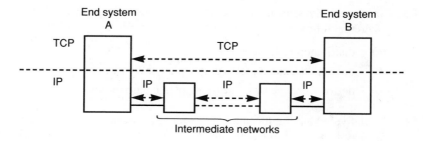

Figure 1.6   TCP/IP operation.

Finally, any discussion on open systems interconnection is incomplete without reference to the Internet and the TCP/IP protocol suite. The US Department of Defense (DoD) had already addressed and produced a solution to internet-working before ISO standardization occurred. The DoD's developments culminated in the Transmission Control Protocol/Internet Protocol (TCP/IP) which is a global internetworking facility available on many interconnected computer networks. Access is easily and inexpensively achieved by subscribing to any of the networks which form the Internet.

TCP/IP assumes that Internet, or some of the underlying networks, are unreliable. With reference to Figure 1.6, reliable operation on an end-to-end basis is achieved by connectionless operation using TCP at layer 4. TCP is therefore provided only at the two end systems rather than in any intermediate network forming part of the connection. This guards against lost packets. IP is implemented at the network layer and is used to route each packet in all intermediate networks. Internet supports e-mail, file transfer and interactive database enquiry services.

Exercises

1.1   Suggest some alternative models to the OSI reference model. Compare and contrast them with the reference model.

1.2   Are there any situations where a reduced version of the reference model may be adequate? Draw such a model.

1.3   Explain the terms 'peer-to-peer communication' and 'encapsulation'. What advantage do these concepts offer designers of open systems?

1.4   Explain what is meant by 'service' and 'protocol' for a given layer of the OSI reference model.

1.5 Summarize the structure of the OSI reference model using a sketch and indicate where the following services are provided:

(a) distributed information services

(b) code-independent message interchange service

(c) network-independent message interchange service

1.6 A relatively long file transfer is to be made between two stations connected to an open system. Assuming that some form of switched communication network is utilized, outline all of the stages that occur to accomplish the file transfer. You should make appropriate reference to each layer of the reference model.

1.7 The OSI reference model is often drawn with the inclusion of an intermediate communications network at the lower three layers. This network does not necessarily have to conform to any particular set of standards, but what is obligatory for inclusion of such networks within an OSI-based design?

Data transmission

2

In Chapter 1 we discussed the nature of data and where it occurs. In this chapter we shall look at the transmission of the data over a transmission channel. First, we shall consider the different techniques used to transmit the data, then the different media over which the data can be transmitted, and finally the different configurations in which communications networks can be arranged.

2.1 Data transmission techniques

Probably the most fundamental aspect of a data communications system is the technique used to transmit the data between two points. The transmission path between the two points is known as a **link** and the portion of the link dedicated to a particular transmission of data is a **channel**.

2.1.1 Communication modes

Communications systems may operate in either a one-way or a two-way mode. In the context of data communications, an example of one-way communication is a **broadcast system** in which data is transmitted from a central station to a number of receive-only stations; there is no requirement for a return channel and therefore no interaction exists. Teletex services are one-way transmission systems. Transmission which is confined to one direction is known as **simplex** operation and is illustrated in Figure 2.1(a).

Simplex operation is limited in terms of its operational capability but it is simple to implement since little is required by way of a protocol. It has a major limitation in that a receiver cannot directly indicate to a transmitter that it is experiencing any difficulty in reception.

Many applications require a channel that allows for communication in both directions for either some form of dialogue or interaction, as is the case with home shopping services or travel agent booking services. Two-way communication is also required if data is to be retransmitted when errors have been detected and some

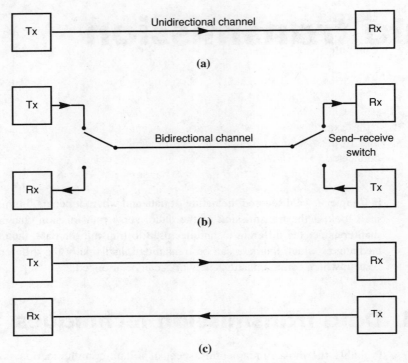

Figure 2.1 Communication modes: (a) simplex; (b) half duplex;
(c) full duplex.

form of repeat request has been sent back to the transmitter. Two possibilities exist
for two-way communication:

(1) **Half-duplex** operation, as shown in Figure 2.1(b), can support two-way
     communication, but only one direction at a time. This is typical of many
     radio systems which, for simplicity and cheapness, employ a common
     channel for both directions of transmission. If a station is transmitting it
     cannot receive from a distant station at the same time. Some form of protocol
     is therefore necessary to ensure that one station is in transmit mode and the
     other is in receive mode at any one time as well as to determine when stations
     should change state.

(2) **Full duplex** operation, as shown in Figure 2.1(c), can support simultaneous
     two-way communication by using two separate channels, one for each
     direction of transmission. This is clearly more costly but is simpler to operate.
     Although at first glance full duplex operation appears to be an obvious
     choice, it must be remembered that two-way transmission is not always
     necessary. Furthermore, where transmission is predominantly in one
     direction, with little data traffic in the reverse direction of transmission, half-
     duplex may be quite adequate.

2.1.2 Parallel and serial transmission

Computer-based systems store and process data in the form of bits arranged in **words** of fixed size. A computer memory typically consists of a series of stores, each of which has a unique address. Computer systems may handle words of 8, 16 or 32 bits. Within many sections of a system data exists and is passed around in parallel form which means that each bit of a word has its own physical conducting path. Examples of **parallel transmission** may be found on printed circuit boards and in printer interface cables. Parallel interfaces, such as a computer interconnected to a printer, need to signal in some way when data is available from the computer and also when the printer is, or is not, ready to receive data. The main reason for such signalling is that very often there are appreciable differences in operating speeds between two interfaced devices. In the case of the computer–printer example, the computer is able to send data at a much faster rate than the printer is able to print. This facility of matching devices is achieved using an exchange of control signals known as **handshaking**. Figure 2.2 shows an example of handshaking in a parallel interface.

Only when the control signal DAV (Data AVailable) changes from low to high does the printer accept whatever data is currently present on the parallel data bus. Clearly the computer must change data on the bus only when DAV is low. Once the computer has signalled that data is available to the printer, the printer

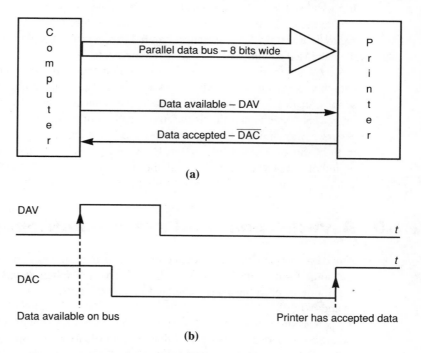

(a)

(b)

Figure 2.2   (a) Parallel transmission and (b) handshaking.

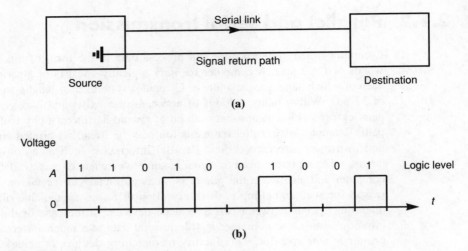

Figure 2.3 Serial transmission.

makes DAC (Data ACcepted) low. When it has completed its printing cycle and is ready for another character, the printer then takes DAC high to indicate to the computer that the data was accepted. Thereafter the sequence may be repeated to print further characters. Parallel transmission is a convenient method of interconnection or interfacing between devices that are situated close together. However, if signals are to be transferred any distance, parallel transmission becomes impractical due to the increased cabling and connector requirements.

An alternative approach is to use **serial transmission** where only a single data line exists and the data bits are sent one after the other along the line. Although serial transmission is inherently slower, the reduced cabling arrangements enable longer connections and this is the usual method of sending data over any distance. Figure 2.3(a) shows a basic serial interface arrangement. Figure 2.3(b) shows data in the form of a serial bit stream. Each bit of this serial data stream occupies a period of time known as a **signal element**. The signals are represented by a positive value (of amplitude A in this case) for a logic 1 and a zero value for a logic 0.

2.1.3 Asynchronous and synchronous operation

Any data communications system comprises as a minimum a transmitter, a receiver and some form of communication channel. The transmitter produces a data stream, the timing of each bit being generated under the control of a **clock** (alternate logic 1s and 0s).

A system in which a transmitter may generate bits at any time employs **asynchronous transmission** and does not necessarily transmit any knowledge of its local clock or bit timing to the receiver. For the receiver to interpret incoming

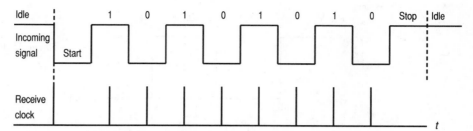

Figure 2.4 Start–stop operation.

signals it must produce a clock of its own which is of the correct frequency and is positioned at, or near, the centre of each received bit in order to interpret each received bit correctly. In practice, an asynchronous receiver does not recover a clock from the incoming signal. Therefore it is generally the case that transmitter and receiver use separate clocks which are of similar frequency by agreeing a nominal transmission rate. To position the receiver clock correctly a method of working known as **start–stop** operation is employed. An example of data transmitted using such a system is shown in Figure 2.4.

When there is no data to send the line remains in an **idle** state. The data is preceded by a start bit which is normally of one-bit duration and must be of opposite sense to that of the idle state. This is followed by several data bits, typically eight. Finally a stop bit is appended which has the same sense as the idle state and which may sometimes be longer than the other bits. A stop bit is included to ensure that the start bit is clearly distinguishable from the last bit of a previous character. When the receiver detects the leading edge of the start bit it starts the production of its receive clock. The first appearance of the clock is so timed to fall at, or near, the centre of the first data bit and is used to **strobe** the bit into a register or latch. Subsequent clock pulses repeat the process for the remainder of the data bits. Clearly, if the arrival rate of the received signal and the local clock are identical, strobing will occur precisely at the centre of the bit durations. In practice, this is seldom the case for, as the number of data bits increases, the coincidence of clock pulses and data bit centres progressively deteriorates. This limits the size of asynchronous signals to a maximum of about 12 bits for reliable detection.

EXAMPLE 2.1

Data is transmitted asynchronously using bits of 20 ms duration. Ignoring any constraints on the number of data bits which may be accommodated between start and stop bits, estimate the maximum number of bits which may be reliably received if the receiver clock is operating at 48 Hz.

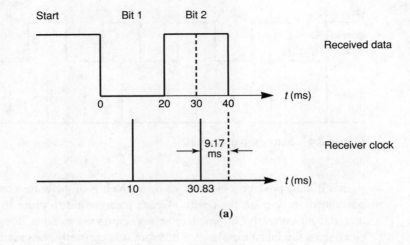

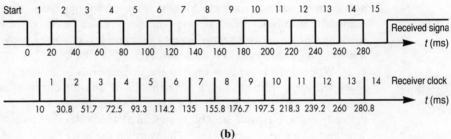

(b)

Figure 2.5 Example 2.1.

Assume that the start bit arranges that the receiver clock starts exactly in the centre of the first received bit period.

> Receiver bit duration $= 20\,\text{ms}$
> Receiver clock period $= 1/48\,\text{Hz}$
> $\approx 20.83\,\text{ms}$

It is clear from Figure 2.5(a) that after the first bit is received, the receiver clock slips by approximately 0.83 ms on each successive bit. The problem therefore resolves into finding how many such slips are necessary to slide beyond half of one received bit period, or 10 ms.

> $10/0.83 = 12$

Therefore the 12th clock pulse after the first one which was perfectly centred on bit 1, that is the 13th actual clock pulse, occurs just on the trailing edge of the 13th bit as may be seen in Figure 2.5(b). This means that bit 13 may or may not be successfully received. The 14th receiver clock pulse occurs 20.83 ms after the 13th clock pulse and is in fact 0.83 ms into the 15th bit. Hence the 14th bit is completely missed. As a result the receiver outputs all of the bits up to and including the 12th correctly, may or may not correctly interpret bit 13 and outputs the 15th bit received as the 14th bit. Hence only 12 bits may be reliably received in this example.

A more efficient means of retaining synchronization in a data communication link is to use **synchronous transmission**. Data which is transmitted synchronously appears at a receiver as a continuous stream of regularly timed bits. The transmitter and receiver must operate in synchronism and this is achieved by using timing signals in the form of a clock. The transmitter generates a clock which can be either transmitted to the receiver over a separate channel or be regenerated at the receiver directly from the transmitted data. For the latter, sufficient timing information must be contained within the transmitted data to enable a component of the transmitter clock frequency to be recovered at the receiver. This clock recovery process is shown in Figure 2.6.

An important factor in ensuring that timing information is contained in transmitted data signals is the choice of signalling technique. This issue is explored further in Chapter 4. In contrast to asynchronous operation, synchronous operation requires a continuous signal to be transmitted, even if no data is present, to ensure that the receiver clock remains in synchronism with that of the transmitter. Furthermore, for a receiver to interpret the incoming bit stream correctly, some **framing** of bits into identifiable fixed-length blocks or frames is necessary at the transmitter. Also, at start-up, it takes quite some time before a stable clock appears at the receiver. Until such time, no data signals may be output from the transmitter. Instead, a special bit pattern, known as a **preamble**, is transmitted to enable the receiver clock to be established. Finally, it is worth noting that the line coding at the transmitter and the clock recovery circuitry at the receiver add a degree of complexity to synchronous transmission which is not required with asynchronous transmission. However, synchronous operation allows much higher data transmission speeds than asynchronous operation and is

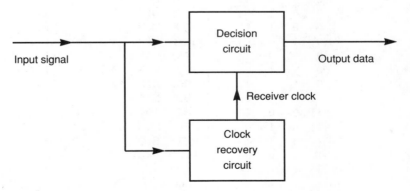

Figure 2.6 Synchronous receiver.

much more efficient, particularly if long frame lengths are used. The efficiency of transmission systems will be dealt with in Chapter 7.

2.1.4 ASCII code

So far, we have looked at the different techniques used to transmit data but not at the way in which the data is represented. Text **characters** are normally represented as fixed-length bit sequences. A number of characters are then grouped together in what is, strictly speaking, an **alphabet** but which is more commonly called a **code**. The alphabet most often used in data communications is the ITU-T alphabet number 5, or more commonly the US national version of this, known as the American Standard Code for Information Interchange (ASCII). Each character in this alphabet is represented by a 7-bit pattern, leading to the 2^7 different characters which are listed in Table 2.1.

The alphabet consists of 96 print characters including both upper- and lower-case letters and digits 0 to 9 ('space' and 'delete' are often included in this grouping) and 32 other characters which cannot be printed but which are associated with control functions. This latter grouping includes 'backspace' and 'carriage return'. A full list of the ASCII control characters is provided in Table 2.2.

If ASCII characters are transmitted asynchronously, start and stop bits are added along with an additional eighth bit which is known as a **parity bit** and is used as a rudimentary check for errors. The parity bit is added so that the transmitted 8-bit

Table 2.1 ASCII code.

Bit			7	0	0	0	0	1	1	1	1
positions			6	0	0	1	1	0	0	1	1
			5	0	1	0	1	0	1	0	1
4	*3*	*2*	*1*								
0	0	0	0	NUL	DLE	SP	0	@	P	\	p
0	0	0	1	SOH	DC1	!	1	A	Q	a	q
0	0	1	0	STX	DC2	"	2	B	R	b	r
0	0	1	1	ETX	DC3	#	3	C	S	c	s
0	1	0	0	EOT	DC4	$	4	D	T	d	t
0	1	0	1	ENQ	NAK	%	5	E	U	e	u
0	1	1	0	ACK	SYN	&	6	F	V	f	v
0	1	1	1	BEL	ETB	'	7	G	W	g	w
1	0	0	0	BS	CAN	(	8	H	X	h	x
1	0	0	1	HT	EM	)	9	I	Y	i	y
1	0	1	0	LF	SUB	*	:	J	Z	j	z
1	0	1	1	VT	ESC	+	;	K	[	k	{
1	1	0	0	FF	FS	,	<	L	\	l	:
1	1	0	1	CR	GS	-	=	M	]	m	}
1	1	1	0	SO	RS	.	>	N	^	n	~
1	1	1	1	SI	US	/	?	O	-	o	DEL

Table 2.2 ASCII control characters.

Character	Meaning	Character	Meaning
NUL	No character	DLE	Data link escape
SOH	Start of heading	DC1	Device control 1
STX	Start of text	DC2	Device control 2
ETX	End of text	DC3	Device control 3
EOT	End of transmission	DC4	Device control 4
ENQ	Enquiry	NAK	Negative acknowledgement
ACK	Acknowlegdement	SYN	Synchronous/idle
BEL	Bell	ETB	End of transmission block
BS	Backspace	CAN	Cancel
HT	Horizontal tabs	EM	End of medium
LF	Line feed	SUB	Substitute
VT	Vertical tabs	ESC	Escape
FF	Form feed	FS	File separator
CR	Carriage return	GS	Group separator
SO	Shift out	RS	Record separator
SI	Shift in	US	Unit separator

sequences contain only an even number of 1s (even parity) or only an odd number of 1s (odd parity). Figure 2.7 shows the ASCII character 'e' as it would typically be transmitted asynchronously using even parity. Note that the stop bit is twice as long as the other bits.

At the receiver a check is made to determine whether the 8-bit sequences still have the same parity. Any character which has not retained its parity is assumed to have been received erroneously. If ASCII characters are transmitted synchronously, a number of characters are normally made into frames.

2.1.5 Signalling rate

Transmission rate, also known as data rate, was defined in Section 1.1 as the number of bits transmitted during a period of time divided by that time and is measured in bits per second (bps). It is important to distinguish between the transmission rate measured in bps and the **signalling rate** or **baud rate** measured in **bauds**. The signalling rate is the rate at which an individual signalling element is

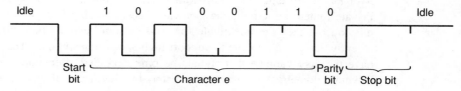

Figure 2.7 ASCII character 'e'.

transmitted and is probably best defined as the inverse of the duration of the shortest signalling element in a transmission. The difference between these two rates is not immediately obvious and is probably best explained by an example.

EXAMPLE 2.2

Asynchronous data is transmitted in the form of characters made up as follows: five information bits of duration 20 ms, a start bit of the same duration as the information bits and a stop bit of duration 30 ms. Determine:

(a)   The transmission rate in bps.

(b)   The signalling rate in bauds.

(a)   The time taken to transmit a single character = $(6 \times 20) + 30 = 150$ ms.
         The number of bits transmitted during this time is 7.
         The transmission rate $= 7/(150 \times 10^{-3}) = 46.67$ bps.

(b)   The shortest signalling element has a duration of 20 ms, therefore the signalling rate $= 1/(20 \times 10^{-3}) = 50$ bauds.

The difference between the bit rate and the baud rate arises because not all of the bits are of the same duration. Another circumstance in which such a difference can arise is if signals are **modulated** onto a **carrier signal** as outlined in the next section.

2.1.6 Modulation and modems

Data is often transmitted using d.c. levels to represent the binary 1s and 0s, in which case it is known as **baseband transmission**, as illustrated in Figure 2.3(b). Such transmission of data requires a channel with a frequency range from 0 Hz to a frequency equivalent to several times that of the baud rate. Until the age of digital public telephone networks, the predominant type of transmission link available was the analogue telephone channel. This was characterized by a bandwidth of 300–3400 Hz or so, and this bandwidth precluded the transmission of d.c. signals. Private rented d.c. circuits have always been available from PTTs for d.c. transmission but they are unsuitable for transmission beyond about 20 km and, until recently, have been unable to support high-speed transmission. These circuits also have the disadvantage that signals become degraded after quite a short distance (several kilometres) and need to be restored to their original shape and amplitude, a process known as **regeneration**. To overcome the speed and distance limitations of

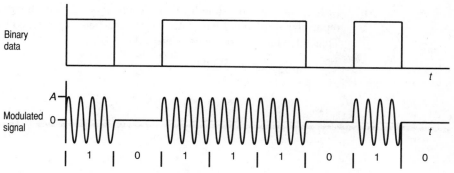

Figure 2.8 Amplitude modulated signals.

d.c. circuits, binary data may be converted into a form which may be readily transmitted over analogue telephone channels. This is achieved by modulating the data onto a voice-frequency carrier signal, as shown in Figure 2.8, for a type of modulation known as **amplitude modulation**.

The resulting modulated signal has a single voice-frequency and two levels of amplitude (0 and *A*), hence the term 'amplitude modulation'. In this way data can be communicated, via the dial-up telephone network, anywhere across the world using a telephone line. The equipment used to produce the modulated signals at one end of a link and convert them back to d.c. at the other end is called a modulator/demodulator or modem. Further modulation techniques will be covered in Chapter 5.

High-speed modems use a multilevel signalling technique to attain high bit rate transmission, using more than one bit per level of a signalling element. For example, if four levels are used then each level can represent two bits, giving the values 00, 01, 10 and 11 to the four levels. Such signalling gives rise to a difference between the bit rate and the baud rate, as Example 2.3 shows.

EXAMPLE 2.3

A modem transmits using an eight-level signalling technique. If each signalling element has a duration of 0.8333 ms, determine:

(a)   the baud rate,

(b)   the bit rate.

(a)   The baud rate is defined as the inverse of the shortest signalling element:

$$\text{Baud rate} = \frac{1}{0.8333 \times 10^{-3}}$$

$$= 1200 \text{ bauds}$$

(b)   If there are eight possible levels for each signal then each signalling element represents three bits (that is, the eight levels are represented by 000 to 111). Thus three bits are transmitted every 0.8333 ms.

$$\text{Bit rate} = \frac{\text{no. of bits}}{\text{time}}$$

$$= 3/0.8333 = 3600\,\text{bps}$$

2.1.7 Bit error rate

In practice, transmission impairments result in data sometimes being received erroneously at the receive end of a data transmission system. Thus a transmitted logic 0 may be received as a logic 1 and vice versa. It is usual to express the number of errors that are likely to occur in a system as a **bit error rate** (BER). A BER of 10^{-4} means that there is a 10^{-4} probability of a bit being received in error. Alternatively we can say that, on average, 1 bit in every 10000 (10^4) will be in error. One of the major causes of error is noise, especially that which is introduced during transmission, which causes the receiver to interpret bits wrongly on occasion. Noise and other transmission impairments, errors and the techniques used to overcome them will be dealt with in detail in Chapter 6. Many systems employ some form of error control to attempt to improve the overall BER. The simplest systems employ some form of **error detection**. In such a system, the receiver may be aware of an error within a group of bits but does not know which particular bit is in error. This system is therefore not able to correct errors. The parity check mentioned in section 2.1.4 is a simple form of error detection which can detect single errors. **Error correction** systems, however, attempt to identify the position of the bits which are in error and hence correct them. In a binary system, correction is effected by a simple change of state of the offending bit (from 0 to 1 or from 1 to 0). No error control systems can guarantee to detect or correct 100% of errors; they are all probabilistic by nature and hence liable to miss some errors. The use of error control only ever serves to improve the BER, although the resulting improvement is often dramatic.

2.2 Transmission media and characteristics

In addition to the transmission techniques used in a data communications system, it is useful to look at the different media over which the data can be transmitted. Media available for transmission of signals are twisted-pair and coaxial-pair metallic conductors and optical fibres. Additionally, signals may be transmitted using radio waves. Twisted-pair cable is used for baseband communication over short distances, typically for the interconnection of peripheral equipment to computer systems. It possesses low inductance but high capacitance which causes substantial attenuation of signals at higher frequencies. Coaxial-pair cable was developed to overcome the deficiencies of twisted-pair conductors by enabling much higher frequencies of operation, typically up to hundreds of megahertz. In addition, its construction produces very little electromagnetic radiation thus limiting any unwanted emissions and reducing the possibility for eavesdropping. Conversely, very little interference may enter coaxial pairs, affording a high degree of immunity to external interference. Coaxial conductors may support data rates in excess of 100 Mbps making them suitable for use in LANs. The velocity of propagation in metallic conductors, twisted-pair and coaxial, is about two-thirds that of the free space velocity of propagation of electromagnetic radiation and is about 200000 km/s.

Radio propagation is difficult to summarize simply within an introductory topic. Its performance is dependent upon the frequency of operation, geographical location and time, and it is a complex subject for design engineers. Radio may be used to good effect in point-to-point operation, especially over difficult or hostile terrain. It is also especially useful for broadcast and essential for mobile applications; there is no need to install many separate physical circuits to link users together. Figure 2.9 illustrates the principal radio propagation methods and gives approximate ranges.

Although not apparent in the figure, it should be added that sky wave propagation may traverse considerably greater distances – many thousands of km – than is indicated, as a result of multiple reflections of the wave from the earth's surface and the ionosphere. There can, in fact, be several such multiple hops. Figure 2.10 relates these propagation methods to the corresponding frequencies, and also makes comparison with those used for metallic and optical media.

A major application of radio transmissions currently attracting attention is the development of the all-digital, GSM pan-European replacement of national analogue cellular mobile radio systems. The most difficult task to be overcome in this application is that of signal fading. This may be due either to a mobile station

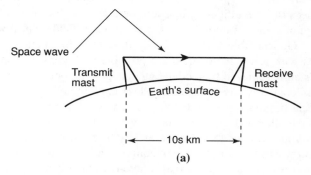

(a)

Figure 2.9   Radio propagation methods: (a) line of sight.

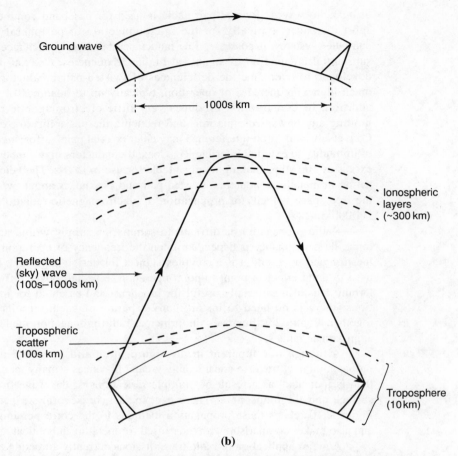

(b)

Figure 2.9 *continued* Radio propagation methods: (b) over the horizon.

moving into a reception area that is weak because of man-made or natural obstacles, or because of multipath propagation effects which change much more rapidly as a result of vehicle motion and hence are more difficult to deal with.

Optical fibre was specifically developed to handle even higher transmission rates with lower losses than are possible with coaxial conductors. Digital transmission is achieved by modulating data onto an optical carrier. There are two main wavelengths where transmission loss is extremely low and which are used with optical fibre systems. As Figure 2.11 shows, one occurs at a wavelength of 1.3 μm and the other at 1.5 μm. The latter wavelength is used in submarine systems where it is important for losses to be as low as possible. In Figure 2.11, losses peak at about 1.4 μm owing to water contamination in the manufacturing process. At shorter wavelengths below 0.8 μm loss rises rapidly as a result of metallic impurities within the glass.

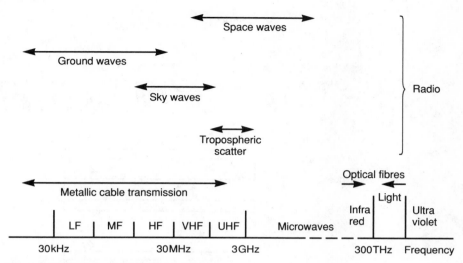

Figure 2.10 Electromagnetic spectrum.

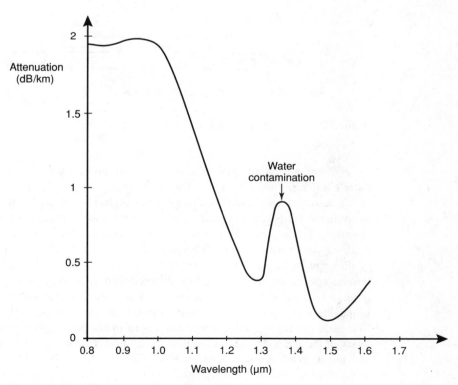

Figure 2.11 Fibre attenuation characteristics.

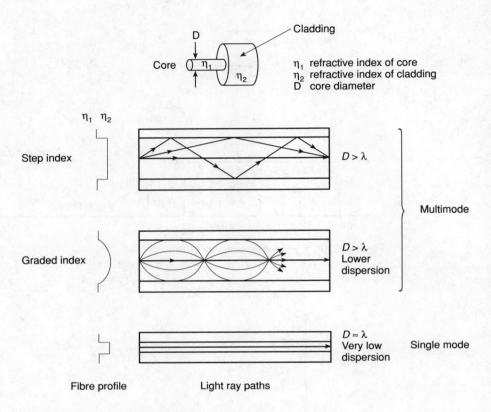

Figure 2.12   Light propagation in optical fibres.

Fibres are constructed of glass or plastic and contain an inner core and an outer cladding. The refractive index of the core is slightly higher than that of the core's surrounding cladding and, as a result, light is contained within the core due to multiple reflections caused by total internal reflection at the core–cladding boundary. Earlier optical communication systems predominantly used fibre with a relatively large core diameter. This permitted individual light rays to follow a range of different paths within the core of the fibre. As a result the rays arrived at the remote end of the fibre at slightly different times, leading to a spreading of the received signal pulses, a phenomenon known as **dispersion**. Such multipath propagation is known as **multimode operation** and is illustrated in Figure 2.12. The earlier systems used a fibre with a **step index** profile in which the refractive index of the glass changed abruptly at the boundary between the core and cladding. The use of **graded index** fibre, in which the refractive index changes gradually, in multimode operation produces an improvement over step index fibre. Here the gradual variation of the refractive index of the core causes the rays of light to follow curved paths, as Figure 2.12 shows. Consequently, the range of path lengths is less than in a step index fibre thus reducing dispersion and allowing higher transmission rates to be used. Transmission rates of up to 100 Mbps and ranges of up to 10 km are possible using multimode, graded index fibre.

Although multimode operation is still used in older systems and also in LANs, **monomode** or **single mode** operation is now dominant for higher speed operation such as high-speed trunk links operated by PTTs. Monomode operation uses a much smaller core diameter and eliminates all but the direct ray of light that can be thought of as passing down the centre of the core thus greatly reducing dispersion and enabling current systems to operate at speeds of 565 Mbps and beyond. Field trials are at present (1996) in operation which will allow rates in excess of 2 Gbps. Lower bit rates have not in the past been economically attractive but as fibre costs fall, this is changing. PTTs are now starting to install fibre for line systems with transmission rates as low as 8 and 4 Mbps.

Optical fibre is potentially much cheaper than coaxial conductors. Optical fibre's inherently low losses enable propagation over several tens of kilometres without signals requiring any regeneration, which is an appreciably greater spacing than is possible using coaxial conductors. Coaxial-pair conductors (and twisted-pairs) suffer from varying degrees of interference, but this is non-existent in fibre systems.

For secure transmission, coaxial or optical media are preferred since they radiate very little, if at all. Radio is inherently insecure and twisted-pair is easily monitored. Where these latter media are unavoidable, data encryption, in which the data is altered so that only the intended recipient should be able to understand it, may be included to add security to the system.

2.3 Network configurations

Having determined the medium to be used in a data communications network, it is necessary to connect communicating stations to the medium. The physical arrangement of a medium used to interconnect stations is known as the **network topology**. Various topologies are shown in Figure 2.13.

One such technology is a fully interconnected **mesh**, shown in Figure 2.13(a). Such a mesh might seem an obvious first approach to interconnecting **nodes**. If there are n nodes, each node requires $n-1$ links to interconnect to every other node. If one assumes that each such link is bidirectional, then the total number of links required in the system may be halved as follows:

$$\text{Total number of links} = \frac{n(n-1)}{2}$$

A little investigation reveals that even for modest values of n, the number of links becomes excessive.

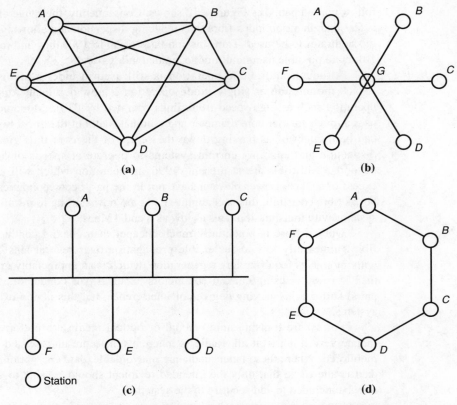

Figure 2.13 Network topologies: (a) fully interconnected mesh; (b) star; (c) bus; (d) ring.

EXAMPLE 2.4

A network is to use a fully interconnected mesh topology to connect 10 nodes together. How many links are required?

Number of nodes, $n = 10$

Total number of links $= \dfrac{n(n-1)}{2}$

$= \dfrac{10 \times 9}{2} = 45$

Even though the size of the network is small, an excessive number of links is necessary. In fact with 10 nodes, the maximum number of links which can be in use simultaneously is only 5. Thus, with 45 links, there is an overprovision by a factor of 9.

An attraction of a mesh configuration is that each station has an equal share in any control strategy. It is able to control the selection of another user directly, a process known as **fully distributed control**, without relying on an alternative switching function. This is in contrast, for example, to a telephone exchange which uses a central switching system. Additionally, stations may be made to perform an intermediate switching function to interconnect two other stations should their direct link fail. This is often referred to as **alternative routing**. Example 2.4 highlights one of the disadvantages of fully interconnected networks, namely excessive link provision except in the case of very small networks. Minimal connectivity can be provided by a tree-like structure exemplified by a **star** configuration as shown in Figure 2.13(b). For a small overhead in switching equipment at the central node, a star network makes vast savings in link provision, as in Example 2.4. Clearly, only 10 links are required, which is more than a fourfold saving. Nevertheless there is still an overprovision by a factor of 2 in that there are twice as many links as are necessary. This control arrangement is called **centralized control**.

The **bus** topology of Figure 2.13(c) is used extensively by LANs. A bus topology may also be thought of as a tree-like structure. Note that the bus may be metallic, fibre or indeed radio. Points to consider are the control of user access to the system and security, since every user potentially 'sees' every message transmitted. Multiple simultaneous accesses to a bus, known as **collisions**, are impossible to avoid with a simple bus and form an important topic of discussion within the section on local area network access control in Section 9.1.

The **ring** topology, as shown in Figure 2.13(d), is also used by LANs and appears as either a true physical ring or a physical bus which is logically ordered as a ring (token bus). Logical rings simply order the passage of data around users in a predetermined and consistent manner. A **token** is normally used to control access to the medium in a ring arrangement, hopefully in a way that is fair to all users. The operation of a **token ring** is as follows. A token in the form of a unique bit-pattern circulates around each station in turn within the ring. A station wishing to transmit seizes the token and transmits its data. When a station has finished with the token, it is released and the next station in the ring receives it and is able to use it if it wishes. Otherwise the token is passed to the next station, and so on. A timer mechanism often exists to ensure that no one station is able to hog the network. Other points such as management of the token and fault conditions will be covered in Chapter 9. Apart from the differences in access control, a bus lends itself more easily to the addition (or subtraction) of stations than does a ring. In some instances these alterations may be done without interruption of service. This is clearly not the case with a ring.

In reality many networks are configured using a combination of both mesh and tree configurations. Figure 2.14 shows the broad framework for the interconnection of telephone exchanges typical of major telephone operators such as British Telecom. Modern telephone exchanges are effectively real-time digital computer systems. Therefore, the switching plan is virtually a network of computer systems, or a data communication networks in its own right. The network shown contains 53 fully interconnected, or meshed, Digital Main Switching Units

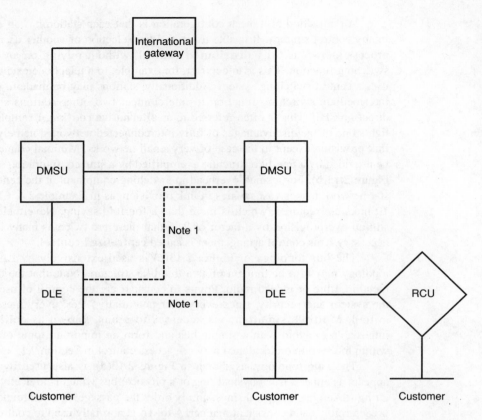

DMSU Digital Main Switching Unit   DLE Digital Local Exchange   RCU Remote Concentrator Unit
Note 1: Supplementary routes subject to demand

Figure 2.14   Telephone network topology.

(DMSUs), each serving a number of Digital Local Exchanges (DLEs) to which customers are connected in star fashion.

Clearly, the choice of topology for a network depends upon factors such as the number of nodes, geographical location and traffic usages and capacities. The debate really focuses upon optimization of the number of switching points which is traded off against link provision. A single star configuration, for instance, does not suit a national network since many of the more distant nodes require long lines to connect to the star. Additionally, if only one hub exists, as is usual in a star configuration, there are an inordinate number of links to deal with at the hub, as well as the associated equipment. An example is if the whole of the United Kingdom were to be served by only one telephone exchange! There are also reliability and security issues to consider and, indeed, duplicate provision may be deliberately built in to deal with such issues.

Exercises

2.1 Explain why a stop bit is necessary in asynchronous transmission.

2.2 With reference to Example 2.2, explain what happens if the first appearance of the clock pulse at the receiver is not exactly in the centre of the first incoming bit.

2.3 Data is transmitted asynchronously at 50 bps. Ignoring any constraints on the number of data bits which may be accommodated between start and stop bits, estimate the maximum number of bits which may be reliably received if the receiver clock is operating at 52 Hz.

2.4 Data is transmitted asynchronously in the form of short blocks consisting of start bit, seven information bits, parity bit and stop bit. All bits are of 9.09 ms duration apart from the stop bit which is of double length. Determine:

(a)   the bit rate

(b)   the baud rate

2.5 A modem operates at a signalling rate of 1200 baud using a 16-level modulation system. Calculate the rate at which data may be transmitted in bits per second.

2.6 Develop an expression to relate the number of bits per symbol to the number of symbols employed in a modulation strategy. Hence relate symbol rate, number of symbols and data rate.

2.7 A computer network is to consist of 20 stations.

(a)   Determine the number of links required if:

(i)   mesh topology is used,

(ii)   star topology is used.

$$L = n(n-1) = \frac{(20)(19)}{2} \quad \frac{380}{2} = 190.$$

20.

(b)   Suppose that 4 stations are designated as hubs, each fully interconnected with the others. The remaining 16 stations are equally distributed to the hubs, any one station only being connected to one hub. Sketch the topology and determine the number of links.

(c)   Compare all three topologies from (a) and (b).

2.8 Explain why synchronous transmission has the potential to operate at higher transmission rates than asynchronous transmission.

Information theory $\boxed{3}$

This chapter introduces the basic aspects of information theory. It establishes the equations for information measure and then develops the equations for the average information content of a code word, that is, the entropy. This then leads to the design of optimum codes and both the Shannon–Fano and the Huffman algorithm are explained.

3.1 The information source

Communications systems are about the transfer of information. The object of any communications system is to transfer a **message** from the source to users at a destination. The message may be such that perfect reproduction at the source is not necessary. For example, consider the following received message:

THA CIT SAT ON THA MIT

Even with the large number of errors in the message most people could probably identify the message as THE CAT SAT ON THE MAT. The message is therefore still understandable to the user at the destination despite the large number of errors. The error probability can be worked out by counting the number of errors and dividing it by the number of **symbols** in the message:

Number of errors = 4
Number of symbols = 17 letters + 5 spaces
$\qquad$ = 22
Error probability for the received message = 4/22

The message is still meaningful because of the inherent redundancy of the English language. The essential property of the message is therefore not necessarily its original set of symbols but something which we refer to as its **information content**. Before the information content of a message can be evaluated we need to establish a means of describing the information source from which the message is generated.

An information source consists of a repertoire of messages from which the desired one is selected. This can be modelled as follows. Assume a source contains a set of symbols denoted by:

$$(x_1, x_2, x_3, ..., x_n)$$

The source's output is a sequence in time of these symbols. The symbols in the source output sequence are a random statistically independent selection from this set and the selections are made according to a fixed probability assignment:

$$P(x_1), P(x_2), ..., P(x_n)$$

Thus a source may be specified by the following:

(1)   the number of symbols in its alphabet denoted by n

(2)   the symbols themselves denoted by $x_1, ..., x_n$

(3)   the probabilities of occurrence of the symbols denoted by $P(x_1), ..., P(x_n)$

A source of this kind is called a **discrete memoryless source**.

Using this model a binary data source may be specified as follows:

(1)   the number of symbols, $n = 2$

(2)   the symbols denoted by $x_1 = 0, x_2 = 1$

(3)   the probability of occurrence of the symbols $P(x_1) = P(x_2) = 0.5$

A telegraph source would be:

$$n = 27 \quad x_1 = A, x_2 = B, ..., x_{26} = Z, x_{27} = \text{space}$$

The probabilities are given by the probability of occurrence of the letters and the space in the English language. This source, however, produces only a first approximation to English, because in English there is a correlation between successive letters. Therefore, a source to produce English needs a knowledge of conditional symbol probabilities as well as their probabilities of occurrence, in other words it needs a memory. This memory is the basis of the redundancy mentioned earlier.

3.2 Information measure

Information received in the form of a message implies that uncertainty existed in the mind of the recipient and that the arrival of the message removed some of the uncertainty. Receipt of a message should, in general, remove some uncertainty. Therefore, a measure of the information content of a message can be based on the amount of uncertainty it removes.

Consider the following three situations:

(1)   An information source consists of the outcomes of tossing a fair coin. The source can be modelled as follows:

The number of symbols, $n = 2$

The symbols represent the outcome of tossing a fair coin so that $x_1 = $ heads, $x_2 = $ tails

The probabilities of occurrence of the symbols are $P(x_1) = P(x_2) = 0.5$

Hence there is equal uncertainty with regard to the outcome of a single toss, so that either outcome removes the same amount of uncertainty and therefore contains the same amount of information.

(2)   When the symbols represent the answers to the question: 'Did you watch television last night?', then the source may be modelled as follows:

The number of symbols, $n = 2$

The symbols are $x_1 = $ Yes, $x_2 = $ No

The probabilities of occurrence of the symbols are $P(x_1) = 0.8$, $P(x_2) = 0.2$ (this assumes that 80% of the population watched television last night)

Hence the receipt of x_1 removes little uncertainty and therefore conveys little information but receipt of x_2 contains a considerable amount of information. It may therefore be deduced that receipt of a high probability symbol contains little information but receipt of a low probability symbol contains a lot more information. This may be rationalized that in the high probability case we could have guessed the outcome but not in the low probability case.

(3)   Consider the source where:

The number of symbols, $n = 2$

The symbols are a binary source with $x_1 = 1$, $x_2 = 0$

The probabilities of occurrence are $P(x_1) = 1$ and $P(x_2) = 0$

The receipt of x_1 is certain and therefore there is no uncertainty and hence no information; x_2 will never be received.

We are now in the position to establish the relationship between the source probability and the information content of a message. Let the information content conveyed by x_i be denoted by $I(x_i)$. Then from the relationships already established we can say that :

If $P(x_i) = P(x_j)$ then $I(x_i) = I(x_j)$

If $P(x_i) < P(x_j)$ then $I(x_i) > I(x_j)$

If $P(x_i) = 1$ then $I(x_i) = 0$

The following function satisfies the above constraints:

$$I(x_i) = \log_b\left(\frac{1}{P(x_i)}\right) = -\log_b(P(x_i))$$

The choice of the logarithmic base b is equivalent to selecting the unit of information. The standard convention of information theory is to take $b = 2$. The unit of information is called the **bit**. The reasons for choosing $b = 2$ is that information is a measure of choice exercised by the source; the simplest choice is that between two equiprobable messages, that is, an unbiased binary choice. The information unit is therefore normalized to this lowest order situation and one bit of information is the amount required or conveyed by the choice between two equally likely possibilities. That is:

If $P(x_i) = P(x_j) = 0.5$
Then $I(x_i) = I(x_j) = \log_2(2) = 1$ bit

Hence in general:

$$I(x_i) = \log_2\left(\frac{1}{p(x_i)}\right) = -\log_2(p(x_i))$$

Unfortunately, logarithms to the base 2 are not usually tabulated, therefore to work out the information content the base of the log must be changed as follows:

$$\log_2 a = \frac{\ln a}{\ln 2}$$

EXAMPLE 3.1

A source consisting of four symbols has the following probabilities of occurrence:

$$P(x_1) = \tfrac{1}{4},\ P(x_2) = \tfrac{1}{8},\ P(x_3) = \tfrac{1}{8},\ P(x_4) = \tfrac{1}{2}$$

Find the information contained by the receipt of the symbol

(a) x_1 (b) x_2 (c) x_3 (d) x_4

The number of symbols, $n = 4$.

(a) $I(x_1) = -\log_2(\tfrac{1}{4}) = 2$ bits
(b) $I(x_2) = -\log_2(\tfrac{1}{8}) = 3$ bits
(c) $I(x_3) = -\log_2(\tfrac{1}{8}) = 3$ bits
(d) $I(x_4) = -\log_2(\tfrac{1}{2}) = 1$ bit

EXAMPLE 3.2

A source consisting of four symbols has the following probabilities of occurrence:

$$P(x_1) = \tfrac{1}{9},\ P(x_2) = \tfrac{2}{9},\ P(x_3) = \tfrac{1}{3},\ P(x_4) = \tfrac{1}{3}$$

Find the information contained by the receipt of the symbol

(a) x_1 (b) x_2 (c) x_3 (d) x_4

The number of symbols, $n = 4$.

(a) $I(x_1) = -\log_2(\tfrac{1}{9}) = \log_2 9$

$$= \frac{\log_e 9}{\log_e 2} = 3.17 \text{ bits}$$

(b) $I(x_2) = -\log_2(\%) = \log_2 4.5$

$\quad = \dfrac{\log_e 4.5}{\log_e 2} = 2.17$ bits

(c) $I(x_3) = -\log_2(\frac{1}{3}) = \log_2 3$

$\quad = \dfrac{\log_e 3}{\log_e 2} = 1.585$ bits

(d) $I(x_4) = 1.585$ bits same as part (c)

The term 'bit' can lead to some confusion, as it is used on the one hand by the communications engineer to represent information content but is also used on the other hand by the microprocessor engineer to indicate the number of binary digits in a word. There is only one condition where the information content of a source in bits is equal to the number of digits in the binary sequence and that is where all the sequences are of equal length and all the symbols are equally probable. This is shown for a binary sequence of length m:

As the symbols are all binary, the number of symbols is specified as $n = 2^m$
The symbols x_i are the binary numbers $i - 1$, for $i = 1, ..., n$
The probabilities of occurrence are $P(x_i) = 2^{-m}$ for $i = 1, ..., n$

Then

$$I(x_i) = \log_2\left(\frac{1}{P(x_i)}\right) = \log_2 2^m = m \text{ bits}$$

This shows that a sequence of m binary digits can convey m bits of information.

For an information source that has symbols $x_i, x_j, x_k, ...$ the probabilities of occurrence of the symbols are $P(x_i), P(x_j), P(x_k), ...$ The information content of receiving a message made up of a number of symbols is simply the addition of the information content of the individual symbols, as follows. When a message $x_i x_j x_k$ is received its probability of occurrence is $P(x_i)P(x_j)P(x_k)$. Hence the information content $I(X)$ is:

$$I(X) = \log_2 \frac{1}{p(x_i)p(x_j)p(x_k)}$$

$$= \log_2\left(\frac{1}{P(x_i)}\right) + \log_2\left(\frac{1}{P(x_j)}\right) + \log_2\left(\frac{1}{P(x_k)}\right)$$

$$= I(x_i) + I(x_j) + I(x_k)$$

3.3 Entropy

In the design and evaluation of a source code it is the average information content of the code that is of interest rather the information content of a particular code word. The average information content conveyed by the symbols of a source code can be shown to be:

$$H(X) = \sum_{i=1}^{i=n} P(x_i)I(x_i) \text{ bits per symbol}$$

EXAMPLE 3.3

A weather information source transmits visibility information with the following probabilities:

Visibility	Probability
Very poor	¼
Poor	⅛
Moderate	⅛
Good	½

Evaluate the entropy of the source code.

Now

$n = 4$
$P(x_1) = \frac{1}{4}, P(x_2) = \frac{1}{8}, P(x_3) = \frac{1}{8}, P(x_4) = \frac{1}{2}$

$$H(X) = \sum_{i=1}^{i=n} P(x_i)\log\left(\frac{1}{P(x_i)}\right) \text{ bits per symbol}$$

$$= P(x_1)\log_2\left(\frac{1}{P(x_1)}\right) + P(x_2)\log_2\left(\frac{1}{P(x_2)}\right) + P(x_3)\log_2\left(\frac{1}{P(x_3)}\right) + P(x_4)\log_2\left(\frac{1}{P(x_4)}\right)$$

$$= \frac{1}{4} \times 2 + \frac{1}{8} \times 3 + \frac{1}{8} \times 3 + \frac{1}{2} \times 1$$

$$= 1\frac{3}{4} \text{ bits per symbol}$$

EXAMPLE 3.4

Determine the entropy for the source with four symbols where all the probabilities are equal to ¼.

The number of symbols, $n = 4$
The symbol probabilities are $P(x_1) = P(x_2) = P(x_3) = P(x_4) = \frac{1}{4}$

Therefore:

$$H(X) = 4 \times \frac{1}{4} \times 2 = 2 \text{ bits per symbol}$$

3.3.1 Maximization of entropy

Examples 3.3 and 3.4 illustrate that sources with the same number of symbols but with different symbol properties have different entropies. For a source of n symbols, maximum uncertainty must exist when all the symbols are equally likely. Hence the maximum information content of a code must occur when all the symbols are equally likely. For n symbols all equally likely, the probability of the occurrence of any one symbol is $1/n$. Therefore the entropy is given by:

$$H(X) = \sum_{i=1}^{i=n} \frac{1}{n} \log_2 n = \log_2 n$$

This is the maximum value of entropy, hence for any source:

$$H(X) \le \log_2 n$$

3.4 Source codes

Coding can serve a wide variety of functions. There are three main types of coding of interest to the communications engineer:

- *Source coding* converts the source signals into a form which minimizes the demand on the transmission channel. It allows the link to carry more traffic and reduces the time taken to send a message. An efficient source code carries the maximum amount of information for the smallest number of bits.

- *Channel coding* is used to detect (and often to correct) errors introduced in the channel. It usually involves introducing redundancy into the code and adds additional bits to the source code.

- *Line codes* are used to match signals to the propagation properties of the channel. This often involves adding timing content to the signal which increases the number of bits required for a particular message.

The best source code is the one which, on average, requires the transmission of the fewest binary digits. This can be quantified by calculating the efficiency of the code. However, before calculating efficiency we need to determine the length of the code. The length of a code is the average length of its code words and is given by:

$$L = \sum_{i=1}^{i=n} P(x_i)l_i$$

where l_i is the number of digits in the ith symbol and n is the number of symbols.

The efficiency of a source code, η, is obtained by dividing the entropy by the average code length:

$$\eta = \frac{H(X)}{L} \times 100\%$$

EXAMPLE 3.5

Evaluate the efficiency for the following source code where:

$n = 4$
$P(x_1) = \frac{1}{4}, P(x_2) = \frac{1}{8}, P(x_3) = \frac{1}{8}, P(x_4) = \frac{1}{2}$
$x_1 = 1, x_2 = 10, x_3 = 100, x_4 = 1000$

The entropy of the source code was previously evaluated in Example 3.3 as $1\frac{3}{4}$ (1.75) bits per symbol.

The length is obtained as follows:

$$L = \sum_{i=1}^{i=n} P(x_i)l_i$$

$$= \frac{1}{4} \times 1 + \frac{1}{8} \times 2 + \frac{1}{8} \times 3 + \frac{1}{2} \times 4$$

$$= 2.75 \text{ digits}$$

Hence the efficiency is:

$$\eta = \frac{H(X)}{L} \times 100\%$$

$$= \frac{1.75}{2.75} \times 100 = 63.6\%$$

Where messages consisting of sequences of symbols from an n-symbol source have to be transmitted to their destination via a binary data transmission channel, each symbol must be coded into a sequence of binary digits at the transmitter to produce a suitable input to the channel.

In order to develop suitable source codes it is useful to prescribe the following descriptors for classifying source codes:

(1)   A **code word** is a sequence of binary digits.

(2)   A code is a mapping of source symbols into code words.

(3)   A code is **distinct** if all its code words are distinct (different).

(4)   A distinct code is **uniquely decodable** if all its code words are identifiable when immersed in any sequence of code words from the code. A code must have this property if its code words are to be uniquely decoded to its original message symbols.

(5)   A uniquely decodable code is **instantaneously decodable** if all its code words are identifiable as soon as their final digits are received.

A uniquely decodable code word is instantaneously decodable if no code word is a **prefix** of another code word. A prefix of a code word is a binary sequence obtained by truncating the code word, for example the code word 1010 has the following prefixes:

1, 10, 101, 1010

It is not absolutely necessary for the code to be instantaneously decodable, but in a real-time system or in any system where speed of operation is vital such a code enables communication to proceed as quickly as possible.

EXAMPLE 3.6

For the source given in Table 3.1, classify each code according to the above definitions and evaluate the length for each instantaneously decodable code.

Table 3.1 Examples of source codes.

Source symbol	$P(x)$	Code 1	Code 2	Code 3	Code 4	Code 5	Code 6
x_1	$5/8$	0	0	00	0	0	0
x_2	$1/4$	1	1	01	01	10	10
x_3	$1/16$	11	00	11	011	110	110
x_4	$1/16$	11	10	10	0111	1110	111

- Code 1 Not distinct, x_3 and x_4 are both coded 11.
- Code 2 Distinct but not uniquely decodable, that is, x_3x_4 is coded 0010 so this could be x_3x_4, or $x_3x_2x_1$, $x_1x_1x_4$ or $x_1x_1x_2x_1$.
- Code 3 Instantaneously decodable, each code word of length 2. $L = 2$ binary digits per code word.
- Code 4 Uniquely decodable since receipt of zero always indicates the start of a new code word. However, it is not instantaneously decodable because the end of the code word cannot be recognized instantaneously.
- Code 5 Instantaneously decodable as receipt of logic 0 always indicates the end of the codeword. L = $1 9/16$ binary digits per code word.
- Code 6 Instantaneously decodable. $L = 1½$ binary digits per code word.

Note: entropy of the source is 1.42 bits per symbol and no uniquely decodable code in the table has a length shorter than this.

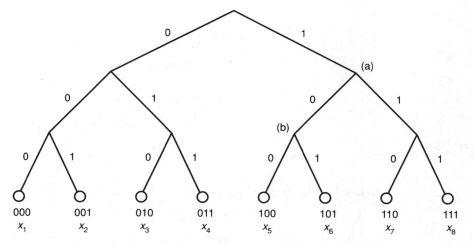

Figure 3.1 Decision tree for a three-digit binary code.

3.4.1 Design of source codes

For a code to be instantaneous, no valid code word must form the first part (the prefix) of another code word. For instance if 11 is a valid code word, then no other valid code word may start with 11.

Instantaneous code words may be generated and decoded by using a **code tree** or **decision tree**, which is rather like a family tree. The root is at the top of the tree and the branches are at the bottom. The tree is read from root to branch. As the binary code is received the digits are followed from the root until the branch is reached when the code word can be read. A code tree is shown in Figure 3.1 for a three-digit binary code. When the code word 101 is received follow the right-hand branch to junction (a) then the left-hand branch to junction (b) then the right-hand branch to the end of the branch to read the code word x_6.

In the case of an instantaneous code, all the code words used to represent the possible states of the source must correspond to the ends of a branch.

A code tree may be used to describe the source code of a weather information system as follows:

Visibility	Probability	Code
Very poor	0.1	111
Poor	0.2	10
Moderate	0.6	0
Good	0.1	110

The corresponding code tree is shown in Figure 3.2.

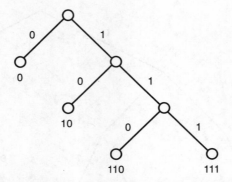

Figure 3.2 Decision (code) tree for a weather information system.

Code trees are very useful for describing a source code once it has been designed, but they are difficult to use in the original design of a code.

If messages are selected from a message set where all messages are equiprobable, then every message carries the same information and direct coding into binary digits is all that is required. For example, if the message set consists of eight symbols, all equally likely, then just a three-digit binary code may be used such that:

$$x_1 = 000 \quad x_2 = 001 \quad x_3 = 010 \quad x_4 = 011$$
$$x_5 = 100 \quad x_6 = 101 \quad x_7 = 110 \quad x_8 = 111$$

However, if the messages have widely different probabilities, then the information associated with the least probable messages is greater then that associated with the most probable messages. Since the least probable messages also occur less frequently, it is sensible that they should be associated with larger groups of binary digits than the most frequent highly probable messages which are associated with the smaller group of binary digits. In this way it is possible to minimize the number of digits required to transmit a message sequence to somewhere near the actual information content of the message. This does, however, mean that variable-length codes will be used, but this is not a problem in most communications systems. Two well-known low-redundancy source coding techniques are the Shannon–Fano and the Huffman codes. These techniques produce optimum codes that are the most efficient codes possible.

3.4.2 Shannon–Fano encoding

First, the messages are ranked in a table in descending order of probability. The table is then progressively divided into subsections of probability near equal as possible. Binary 0 is attributed to the upper subsection and binary 1 to the lower subsection. This process continues until it is impossible to divide any further. This method ensures that the most probable messages have fewer bits than the less probable ones. It also ensures that no code word forms the initial part of another code word hence it is instantaneously decodable.

Consider the following 8-symbol source:

$n = 8$

$P(x_1) = P(x_2) = 0.25$, $P(x_3) = 0.15$, $P(x_4) = 0.1$, $P(x_5) = 0.1$, $P(x_6) = 0.08$, $P(x_7) = 0.06$, $P(x_8) = 0.01$

The Shannon–Fano code of this source is obtained by performing the following (see Table 3.2):

(1) List the symbols in descending order of the probabilities.

(2) Divide the table into as near as possible two equal values of probability. *In this case the table will be divided into two just after x_2.*

(3) Allocate binary 0 to the upper section and binary 1 to the lower section.

(4) Divide both the upper section and the lower section into two. *In this case the upper section will be divided just after x_1 and the lower section just after x_4.*

(5) Allocate binary 0 to the top half of each section and binary 1 to the lower half.

(6) Repeat steps (4) and (5) until it is not possible to go any further.

Table 3.2   Shannon–Fano encoding.

Source message	Probability	Code word
x_1	0.25	00
x_2	0.25	01
x_3	0.15	100
x_4	0.1	101
x_5	0.1	110
x_6	0.08	1110
x_7	0.06	11110
x_8	0.01	11111

Length, $L = 2.72$ digits per symbol
Entropy, $H(X) = 2.67$ bits per symbol
Efficiency, $\eta = 98.2\%$

EXAMPLE 3.7

Design the Shannon–Fano code for the following source:

$n = 6$

$P(x_1) = 0.4$, $P(x_2) = 0.2$, $P(x_3) = 0.15$, $P(x_4) = 0.15$, $P(x_5) = 0.05$, $P(x_6) = 0.05$

The Shannon–Fano coding is shown in Table 3.3.

Table 3.3 Shannon–Fano coding for Example 3.7.

Source message	Probability	Code word
x_1	0.4	00
x_2	0.2	100
x_3	0.15	101
x_4	0.15	110
x_5	0.05	1110
x_6	0.05	1111

Length, $L = 2.3$ digits per symbol
Entropy, $H(X) = 2.25$ bits per symbol
Efficiency, $\eta = 97.8\%$

3.4.3 The Huffman code

This is a graphical technique which produces an optimum code similar to that produced by the Shannon–Fano method. The code words are listed in descending order of their probabilities. The lowest two probabilities are then combined to give a new probability (the sum of the two). The procedure is then repeated until all the message groups have been combined and the sum of the final probability is 1. At each combination point the binary symbols 0 and 1 are allocated, with binary 0 being associated to the upper division and binary 1 to the lower one. The diagram is then read from right to left.

Consider the 8-symbol source used for the Shannon–Fano method:

$n = 8$
$P(x_1) = P(x_2) = 0.25$, $P(x_3) = 0.15$, $P(x_4) = 0.1$, $P(x_5) = 0.1$, $P(x_6) = 0.08$,
$P(x_7) = 0.06$, $P(x_8) = 0.01$

The probabilities are ranked in descending order with x_1 at the top and x_8 at the bottom as shown in Figure 3.3. The two lowest probabilities $P(x_8)$ and $P(x_7)$ are combined to give a new probability $P(x_7') = 0.07$. The two lowest probabilities $P(x_6)$ and $P(x_7')$ are then combined to give a new probability $P(x_6') = 0.15$. The two lowest probabilities $P(x_4)$ and $P(x_5)$ are then combined to give a new probability $P(x_4') = 0.2$. The two lowest probabilities $P(x_3)$ and $P(x_6')$ are then combined to give a new probability $P(x_3') = 0.3$. The two lowest probabilities $P(x_2)$ and $P(x_4')$ are then combined to give a new probability $P(x_2') = 0.45$. The two lowest

Source message

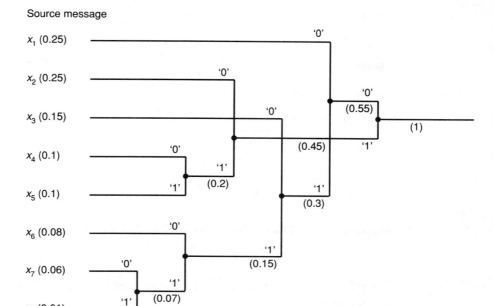

Figure 3.3 Huffman code.

probabilities $P(x_1)$ and $P(x_3')$ are then combined to give a new probability $P(x_1') = 0.55$. The final two probabilities $P(x_1')$ and $P(x_2')$ are combined to give the resultant overall probability of 1. Binary 1 and binary 0 are then allocated at every junction and the diagram is read from right to left.

$$x_1 = 00, x_2 = 10, x_3 = 010, x_4 = 110, x_5 = 111, x_6 = 0110, x_7 = 01110, x_8 = 01111$$

The efficiency of this code is the same as that obtained from the Shannon–Fano code and, in general, both methods always give the same efficiency.

EXAMPLE 3.8

Use the Huffman method to obtain the optimum code for Example 3.7 where:

$n = 6$

$P(x_1) = 0.4, P(x_2) = 0.2, P(x_3) = 0.15, P(x_4) = 0.15, P(x_5) = 0.05, P(x_6) = 0.05$

The Huffman code is constructed as shown in Figure 3.4.

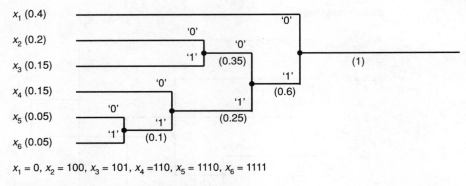

$x_1 = 0$, $x_2 = 100$, $x_3 = 101$, $x_4 = 110$, $x_5 = 1110$, $x_6 = 1111$

Figure 3.4 Huffman code for Example 3.8.

Practical example of the implementation of a Huffman code

To send a page of information a fax machine scans an A4 page then codes the information into a binary code suitable for transmission. Fax machines are classified by group numbers as shown in Table 3.4. Huffman codes are used in group 3 facsimile (fax) systems to increase the speed at which pages of information may be sent.

Group 1 and 2 terminals are virtually obsolete in that they are now no longer produced. Group 3 terminals are the dominant ones. The method of operation of a group 3 fax is as follows. The A4 page of information is scanned from the top left-hand corner to the bottom right-hand corner. Each line is subdivided into 1728 picture elements (pels). Each pel is quantized into either black or white. In the vertical direction the page is scanned to give approximately 1145 lines. Hence the total number of pels per page is just under 2 million. If a binary 1 is used to represent a white pel, and a binary 0 is used to represent a black pel this produces just under 2 million bits per page. The signalling rate specified for group 3 fax terminals is 4800 bps, hence if the above scheme is used, the time taken to transmit a page is:

$$2 \times 10^6/4800 = 417 \text{ seconds (approximately 7 minutes)}$$

Table 3.4 Facsimile machine classifications.

Group number	Transmission speed of an A4 page
1	6 minutes
2	3 minutes
3	typically under 1 minute

Table 3.5 Typical Huffman codes used in group 3 fax terminals.

Number of consecutive pels	Code allocated
7 white	1111
6 white	111111
5 white	1100
3 white	1000
2 white	0111
8 black	000101
4 black	011
3 black	10
2 black	11
1 black	010

The scanned data contains considerable redundancy and by suitable signal processing it is possible to reduce the transmission speed by a factor of about 10. The redundancy reduction method operates as follows. Instead of sending a binary digit for each pel a binary code is produced which indicates the length of each run of black or white pels. The allocation of codes to each run length is achieved by using the Huffman algorithm, based on the probability of occurrence of a particular run length. The Huffman algorithm used for group 3 fax terminals produces the codes shown in Table 3.5.

The Huffman algorithm helps to reduce the time to send an A4 page of information by a factor of approximately 10.

Exercises

3.1 State the maximum entropy of a 16-symbol source.

3.2 Determine which of the following codes are distinct, uniquely decodable and instantaneously decodable.

(a) $x_1 = 0$, $x_2 = 11$, $x_3 = 111$

(b) $x_1 = 0$, $x_2 = 10$, $x_3 = 111$, $x_4 = 110$

(c) $x_1 = 11$, $x_2 = 10$, $x_3 = 01$, $x_4 = 001$, $x_5 = 000$

3.3 Briefly explain the functions of

(a) source coding

(b) channel coding

(c) line coding

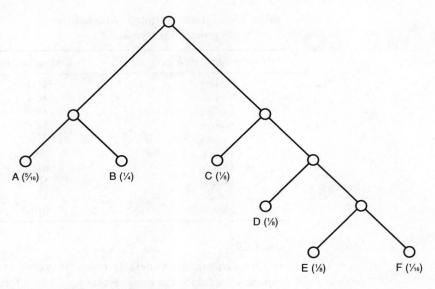

Figure 3.5   Coding tree for question 3.5.

3.4   For each of the following sources obtain an optimum source code, then calculate its entropy, length and efficiency.

(a) $n = 5$, $P(x_1) = 0.5$, $P(x_2) = P(x_3) = P(x_4) = P(x_5) = 0.125$

(b) $n = 4$, $P(x_1) = 0.35$, $P(x_2) = 0.3$, $P(x_3) = 0.2$, $P(x_4) = 0.15$

(c) $n = 4$, $P(x_1) = 0.5$, $P(x_2) = 0.25$, $P(x_3) = 0.1875$, $P(x_4) = 0.0625$

3.5   Figure 3.5 shows the coding tree for a signal source with six discrete states, together with the probabilities for each state (shown in parentheses). Use the tree to obtain the code for each state and evaluate the efficiency of the code.

3.6   For the signal source of Exercise 3.5, design an optimum code and compare its efficiency with that obtained from the coding tree.

Line codes

4

This chapter describes the line codes that are used in present-day systems. It establishes the reasons for using line codes and then looks at the advantages and disadvantages of common codes such as AMI, CMI and HDB3. The particular advantages of block codes are looked at and the chapter concludes by investigating scramblers.

4.1 Line code characteristics

A line code is designed to match the transmitted data to the characteristics of the line over which it is to be transmitted. For baseband circuits such as metal cables it is normally the last coding function carried out before transmission. In the case of radio links, optical fibre and other systems in which modulation takes place, the line coding either takes place prior to modulation or is incorporated into the modulation process. Chapter 5 describes the various modulation schemes that are in common usage.

The main characteristics of line codes are as follows:

(1) Sufficient timing information must be included in the data for the remote receiver to be able to extract a clock signal from the data. This can be easily achieved if the line signalling has frequent transitions at the clock frequency. Long sequences of binary 1s or 0s cause loss of synchronization and must therefore be avoided.

(2) The spectrum of the transmitted data needs to be appropriate to the line characteristics. Lines with coupling capacitors or transformers do not pass d.c. components and therefore nominally constant voltage levels can drift up and down as a result of capacitor charging or discharging. This results in a wandering of the signal levels, known as baseline wander, and is extremely undesirable. In order to minimize this effect the spectrum of the line code should not include a d.c. component.

(3) The transmitted data also needs to be compatible with the bandwidth available for transmission. A reduction in the transmitted symbol rate and

hence the signal bandwidth can be achieved by using multilevel symbols so that each symbol represents more than one bit. Such systems are, however, more susceptible to noise and require a higher signal-to-noise ratio than a straightforward binary system.

4.2 Return-to-zero and non-return-to-zero signalling

The simplest way to transmit data signals is to represent the two binary digits by two different voltage levels. If such a system has a constant voltage level during a bit interval then it is known as **non-return-to-zero** (NRZ). A positive voltage may be used to represent binary 1 and a negative voltage to represent binary 0 (Figure 4.1). For this code a long string of binary 1s or 0s produces baseline wander and, because of a lack of transitions, may also result in a loss of synchronization.

NRZ codes are the easiest to implement and also make efficient use of bandwidth. However, they have severe limitations in that they lack synchronization capability and have a d.c. offset giving rise to baseline wander. NRZ codes are commonly used for digital recording but, because of their limitations they are seldom used for signal transmission applications.

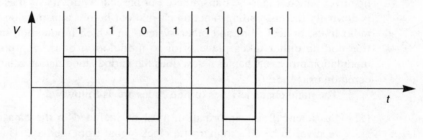

Figure 4.1 NRZ digital signals.

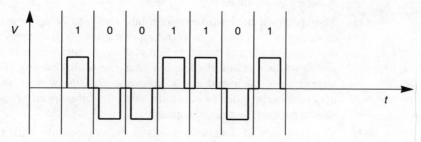

Figure 4.2 RZ digital signals.

An alternative technique used to overcome some of the above difficulties is **return-to-zero** (RZ). In this method binary 1 is represented by a positive pulse and binary 0 by a negative pulse and the waveform returns to zero after each encoded bit (positive or negative). Figure 4.2 shows an encoded sequence in this format and it can be seen that three distinct amplitude levels are now required. This method guarantees a large number of transitions and good clock synchronization can be ensured. However, if a long string of binary 1s or 0s are being transmitted, d.c. offsets causing baseline wander will occur.

4.3 Bipolar alternate mark inversion

For **Bipolar Alternate Mark Inversion** (AMI) a binary 0 is represented by zero line voltage and binary 1 is represented by a positive or negative alternate pulse. The successive binary 1 pulses alternate in polarity as shown in Figure 4.3. AMI derives from the long-established terms 'mark' and 'space' which are used in telegraph systems to indicate binary 1 and binary 0, respectively. AMI has several advantages over the previous codes described. Good synchronization is maintained providing plenty of binary 1s are to be encoded. As the encoded binary 1s alternate in polarity there is no d.c. offset and therefore no problems with baseline wander. The bandwidth requirement is also a lot less than that required for RZ systems. However, a problem still remains in that there may be a loss of synchronization when long strings of binary 0s are to be encoded.

4.4 Code radix, redundancy and efficiency

In order to facilitate our discussion of line codes it is useful to specify a number of common definitions with regard to efficiency and redundancy. A code **radix** is defined as the number of different signalling states used. For baseband systems it is

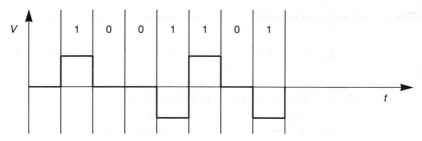

Figure 4.3   Bipolar AMI.

the number of different voltage levels used by the symbols. For example, AMI has radix of 3 (ternary code).

The efficiency of the code is defined as the ratio of the information per symbol used to the information per symbol available. Let I_u be the information per symbol used and I_a be the information per symbol available. Then:

$$\text{Efficiency} = \eta = \frac{I_u}{I_a}$$

Redundancy in line codes may be defined as:

$$\text{Redundancy} = \frac{I_a - I_u}{I_u}$$

$$= \frac{1}{\eta} - 1$$

EXAMPLE 4.1

Evaluate the efficiency and redundancy for an AMI code.

AMI has three signalling levels so the radix is equal to 3. As the three signalling levels are available for sending information the probability of any level being available is ⅓ hence the information available is:

$$I_a = \log_2 3 = 1.58 \text{ bits per symbol}$$

However, in AMI there are really only two signalling levels used, as binary 1 is represented by one of two different levels hence the information used is:

$$I_u = \log_2 2 = 1 \text{ bit per symbol}$$

Hence:

$$\text{Efficiency} = \eta = \frac{1}{1.58} = 0.63 \text{ or } 63\%$$

$$\text{Redundancy} = \frac{1}{0.63} - 1 = 0.59 \text{ or } 59\%$$

The significance of the efficiency of a code is that, for a given code radix and signalling rate, a more efficient code can convey a greater information rate. This implies that if a link is required to convey a given information rate then, for a given radix, a more efficient code will require a lower signalling rate. A lower signalling rate means better use of the available spectrum and allows longer cable lengths between repeaters. Efficiency and redundancy relate only to the information-carrying capacity of a code and give no information as to other important characteristics such as synchronization and baseline wander.

4.5 Codes in the ITU-T digital transmission hierarchy

ITU-T specifies two digital transmission hierarchies, one based on 2048 kbps and the other on 1544 kbps. The former is used in Europe and is probably the most widely adopted around the world. The latter is used in the United States and Japan and other countries which have been influenced by them. Table 4.1 lists the data rates and the corresponding line codes.

Table 4.1 ITU-T digital transmission hierarchy.

Data rate (kbps)	Code
2048	HDB3
8448	HDB3
34368	HDB3
139264	CMI
565000	CMI
1544	AMI with scrambling or restricted data, or B8ZS
6312	B6ZS or B8ZS
32064	Scrambled AMI
44736	B3ZS

The most common codes in Europe are HDB3 and CMI, whereas the United States uses AMI and a slight variation of HDB3 called BnZS which will be explained in Section 4.8.

4.6 High density bipolar

High Density Bipolar (HDB3) is a modified form of AMI which solves the problem of loss of synchronization associated with long strings of zeros. If a string of four zeros is to be transmitted, the last one of the zeros is represented by a **violation pulse** that has polarity that violates the alternate mark inversion (AMI) coding rule, as shown in Figure 4.4.

This first violation pulse is easily identified at the receiver since it violates the AMI coding rule. However, this modification to AMI is liable to build up a d.c. component and a further modification is used to prevent this. An extra pulse called the **balancing (B) pulse** is inserted in place of the first pulse in a group of 4 zeros whenever necessary to prevent two successive violation pulses having the same polarity. For example, Figure 4.5 shows the data sequence

101000010100001010000101

encoded into HDB3.

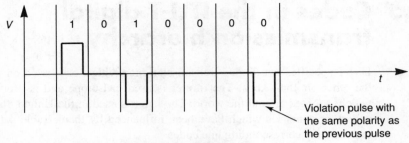

Figure 4.4 Violation pulse in HDB3 coding.

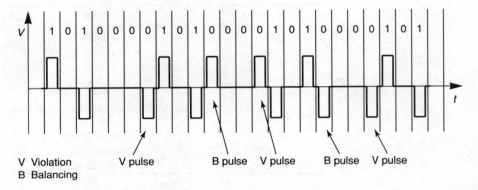

V Violation
B Balancing

Figure 4.5 HDB3 encoded data stream.

The HDB3 coding rules mean that both the encoder and the decoder need to have a memory incorporated in them. The encoder needs to store the polarity of the previous violation pulse and needs to check if the next violation pulse has the same polarity, in which case a balancing pulse needs to be inserted. Hence each input bit has to be stored by the encoder until three subsequent bits have arrived; at that time a decision can be made about inserting a balancing pulse or not. The decoder needs to store the polarity of the received pulses to see if a violation has occurred. It recognizes a balancing pulse by noting that whenever a violation occurs after only two zeros, the previous pulse must be a balancing pulse.

4.7 Coded mark inversion

Coded Mark Inversion (CMI) is used at high signalling rates in preference to HDB3 as it uses simpler circuitry for encoders and decoders. This is a big advantage as high-frequency circuits tend to be complex and expensive. CMI is a binary code, but can also be seen as a modification to AMI designed to remove strings of zeros. Binary 1s are represented by an alternating polarity as for AMI using

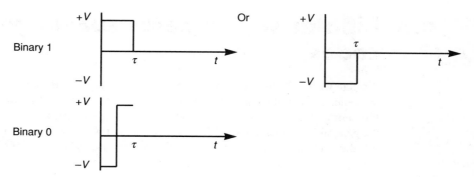

Figure 4.6 CMI coding rules.

full-width NRZ pulses. Binary 0s are represented by a half period at a negative voltage followed by a half period at a positive voltage. These encoding rules ensure a transition on every bit which aids synchronization and clock recovery. There is no d.c. offset and hence baseline wander is not a problem. Figure 4.6 illustrates the coding rules for CMI.

EXAMPLE 4.2

Using the CMI encoding rules encode the data stream 1100101.

Figure 4.7 shows the two possible versions of encoding the data stream 1100101.

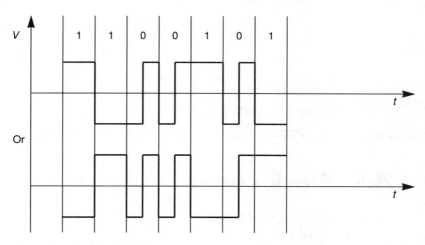

Figure 4.7 CMI encoded waveform.

4.8 Bipolar with *n* zeros substituted codes

Bipolar with *n* Zeros substituted codes are normally called B*n*ZS codes. B*n*ZS is the generic name of a class of codes of which HDB3 is the best-known example. In this terminology HDB3 is B4ZS, indicating that the fourth zero is substituted. In the general implementation of this code the same rules as HDB3 apply with regard to balancing pulses but the violation pulse is now inserted in place of the *n*th zero.

EXAMPLE 4.3

Encode the data stream 1010000001100000011 into B6ZS format assuming that the previous violation and mark were both of negative polarity.

For a B6ZS code a violation pulse is inserted in place of the 6th zero and the balancing pulse is inserted to prevent two consecutive violations in the same direction. Figure 4.8 shows the encoded data for B6ZS.

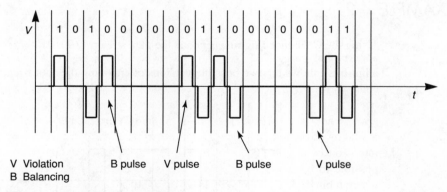

V Violation
B Balancing

Figure 4.8 B6ZS coded data.

4.9 Block codes

All the line codes looked at so far are examples of **bit-oriented codes** where the coding rules operate on individual bits. In **block codes** the coding rules operate on a block of bits which means that they often require more complex encoding and decoding circuitry. Because of their high efficency, block codes allow longer cable lengths between repeaters as explained in Section 4.4. They therefore have

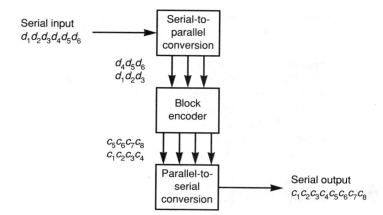

Serial input
$d_1d_2d_3d_4d_5d_6$

Serial-to-parallel conversion

$d_4d_5d_6$
$d_1d_2d_3$

Block encoder

$c_5c_6c_7c_8$
$c_1c_2c_3c_4$

Parallel-to-serial conversion

Serial output
$c_1c_2c_3c_4c_5c_6c_7c_8$

Figure 4.9   Block code encoder.

applications in long-distance transmission systems, where the extra complexity in the encoding and decoding circuits is justified by the saving in the number of repeaters. Like the codes previously described they are designed to prevent baseline wander and to have good timing content.

The basic principle behind the design of block codes is that blocks of input bits are encoded to give blocks of code symbols. This is different to all the codes looked at so far which involve coding a single bit to give one or more code symbols.

The encoder for a block code first has to break the incoming data stream into appropriately sized blocks (sometimes called words). It then encodes each block of input bits into a block of code symbols (code words) Finally, it has to transmit the code words serially. Figure 4.9 is a typical block diagram for a block encoder.

A binary block code can be specified as an nBmB code, where n binary bits are encoded into m binary bits. So in the case of a 3B4B code the serial input data is first of all broken down into blocks 3 bits wide, the 3-bit blocks are then encoded into 4-bit blocks and the 4-bit blocks are then transmitted as serial data. It is also possible to have a ternary block code which can be specified as an nBmT code, where n binary bits are encoded into m ternary symbols.

It is not unusual for the incoming data to have no inherent block structure; the position of the initial block boundary chosen by the encoder may therefore be arbitrary. However, once the initial block boundary is chosen, the encoder can then take successive sequences of n adjacent symbols.

One of the prime advantages of block codes is that they can be designed to have an efficiency greater than either HDB3 or AMI. The nature of a block code is such that a block of m binary output bits from an encoder is carrying the information from n binary input bits. The information per symbol used is n/m bits and the information per symbol available for a binary symbol is 1 bit. Consequently, the efficiency of a general nBmB code is:

$$\eta = \frac{n}{m} \times 100\%$$

A code with $n > m$ implies an efficiency greater than 100%, which is impossible because the coded blocks would need to carry more information than is theoretically possible. Codes with $m = n + 1$ are particularly common since they have higher efficiency than any with $m > n + 1$.

The efficiency of a block code is determined by the choice of n and m. In general, as $2^m > 2^n$ for a binary code or $3^m > 2^n$ for a ternary code, not all the possible code words are required by the code. This leaves some flexibility in the choice of which code words are used. It is this flexibility that allows block codes to be designed with such good timing content. It also means that block codes can be designed to suit a range of applications.

4.9.1 3B4B block code

The 3B4B block code is one of the simplest types of block code, but it does illustrate all of the characteristics of a block code and will therefore be described in some detail.

Block codes are fully described in a **coding table** which shows how each input word is coded. Table 4.2 gives all the encoding rules for the 3B4B code. The table assumes that the code is implemented by positive and negative pulses where + indicates a positive pulse and – indicates a negative pulse.

Table 4.2 has been constructed to ensure that the encoded words have the following properties:

● There is no mean voltage offset (to prevent baseline wander).
● There are frequent transitions to maintain timing.

The left-hand column lists all the possible input 3-bit binary words. The next three columns contain all the possible output words according to the coding rules. The fifth column contains the code word disparity. The disparity, or digital sum, of a code word is a count of the difference between the number of positive and negative pulses and is therefore a normalized measure of the mean voltage of the

Table 4.2 3B4B coding table.

Input	Output			Disparity
	Negative	0	Positive	
001		– – + +		0
010		– + – +		0
100		+ – – +		0
011		– + + –		0
101		+ – + –		0
110		+ + – –		0
000	– – + –		+ + – +	±2
111	– + – –		+ – + +	±2

Table 4.3 Example of a running digital sum.

Coded word	Running digital sum
$- - + +$	0
$- - + -$	-2
$+ - - +$	-2
$+ + - -$	-2
$+ - + +$	0
$- + + -$	0

word. A word with equal numbers of positive and negative pulses has zero disparity and a zero mean voltage. A word such as this is described as a **balanced word**. **Unbalanced words** can have positive or negative disparity. For example the code word $+ + - +$ has a disparity of +2. To ensure that there is no mean voltage offset, balanced code words are used in Table 4.2 for the output where possible, that is, for input words 001 to 110. The remaining two input words have to be encoded to unbalanced words. To maintain the zero mean offset, each of these two input words can be encoded to a word with either a positive mean voltage or a negative mean voltage. The choice between the two for a particular input word is made at the time of encoding and depends on the value of the **running digital sum**. During encoding the encoder keeps a record of the running digital sum of each transmitted word. Table 4.3 shows how the value of the running digital sum changes during the transmission of some coded words.

The encoder consults the value of the running digital sum whenever an unbalanced word must be transmitted. If the sum is negative or zero, the positive disparity option is selected, if it is positive the negative disparity option is selected. When the input word is encoded by a balanced word then no choice needs to be made and the running digital sum is unchanged.

Following the above encoding rule and assuming that the register initially contains 0, the running digital sum is always either 0 or -2 between the transmission of words. So the encoder can be regarded as having two different states, defined by the current running digital sum.

EXAMPLE 4.4

For the 3B4B code just described and assuming that the encoder starts with the running digital sum equal to zero, encode the data sequence 000 001 111 101 111, specifying the value of the running digital sum after each word is encoded.

The encoding is shown in Table 4.4.

Table 4.4 Answer to Example 4.4.

Input word	Running digital sum before encoding	Encoded word	Running digital sum after encoding
000	0	− − + −	−2
001	−2	− − + +	−2
111	−2	+ − + +	0
101	0	+ − + −	0
111	0	− + − −	−2

The output sequence is therefore:

$$-- + --- + + + - + + + - + -- + --$$

In order to decode a 3B4B code the decoder needs to break the incoming serial data sequence into 4-bit words. The decoder must know where to break the serial data into words to get the correct blocks of code symbols; this is known as block alignment. Once the alignment has been carried out, the decoding may be completed using a look-up table which consists of all the possible binary forms A decoding table is shown in Table 4.5.

Code words marked with an asterisk are called **forbidden words**; these are words which are not used in the coding table. However, they might appear in the coded data arriving at a decoder as a result of errors in the received data. The

Table 4.5 Decoding for 3B4B codes.

Received code word	Decoded to
− − − − *	001
− − − + *	000
− − + −	000
− − + +	001
− + − −	111
− + − +	010
− + + −	011
− + + + *	011
+ − − − *	100
+ − − +	100
+ − + −	101
+ − + +	111
+ + − −	110
+ + − +	000
+ + + − *	111
+ + + + *	110

decoder needs to know how to decode them, therefore an entry must be included in the decoding table. The forbidden words are decided on the basis of producing the lowest average number of errors in the output.

4.9.2 Error detection

As noise is present in all transmission systems it is more than likely that errors will occur in a received data stream. Line codes use a particular coding rule that may be used as a means of error detection. The received data stream is checked to see if it is consistent with the particular rule. Consider, say, an AMI code that is used to encode the data 1011001. This produces the code:

$$+ 0 - + 0\ 0 -$$

Now suppose an error were to occur in the fifth symbol such that the received code became:

$$+ 0 - + - 0 -$$

A violation has occurred in the coding rule and an error can be detected. In this system we have error detection but not error correction, and although single errors can be detected multiple errors may be missed.

In block codes, the Digital Sum Variation (DSV) can be used to detect errors. As the running digital sum can vary between 0 and –2 between words and by ±2 during words its total variation can be from –4 to +2. Therefore, the total DSV is 6 with bounds of – 4 and +2. A single error will eventually cause the measured running digital sum to exceed its bounds. Decoders can monitor the DSV bounds and if they are exceeded an error can be indicated.

Figure 4.10 shows how the DSV varies on the receipt of a particular data stream. An example is also shown in Figure 4.10 of the effect of a single error. For this situation the DSV exceeds its bounds and therefore an error can be detected. Again, this system gives error detection, not error correction.

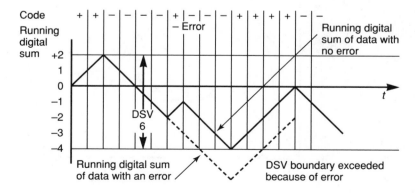

Figure 4.10 Digital sum variation.

4.10 Frequency spectra characteristics of common line codes

In Section 4.1 we said that one of the main characteristic of a line code is that the spectra of the transmitted data should be appropriate to the line characteristics. We therefore need to examine the spectrum of the various line codes we have looked at in order to determine therir suitability for transmission. In comparing line codes we need to take care when evaluating the spectral characteristics as, for most cases, the spectral diagram of each code is dominated by a particular pulse shape. Figure 4.11 shows the frequency characteristics of the line codes already considered. The graphs are obtained by using impulses as digits. The spectrum of randomly encoded data using such impulses is constant (flat). Hence for any such data encoded by the normal coding rules the deviation from a flat spectrum is due entirely to the code.

The spectral characteristics of the codes can now be compared using the information contained in Figure 4.11. For the straightforward binary code the spectrum is flat and includes d.c. and low-frequency components thereby giving rise to baseline wander.

AMI code has some very attractive properties; the spectral diagram shows no d.c. component and very little low-frequency content thereby reducing considerably the effects of baseline wander. The bandwidth required is equal to the transmission rate. It is a very simple code to implement but does have the big disadvantage of poor timing content associated with long runs of binary zeros.

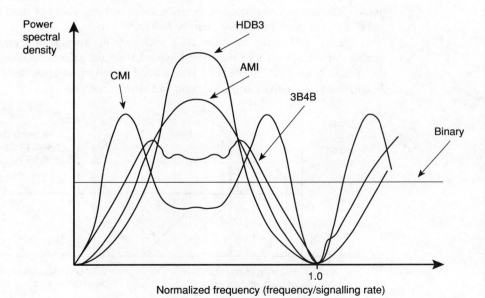

Figure 4.11 Frequency spectral characteristics of some line codes.

HDB3 has similar characteristics to AMI but has good timing content. Its main disadvantage, as already mentioned, is that it is more complex to implement and this means a bigger cost, particularly at high data rates.

CMI has a power spectrum that is fairly evenly distributed throughout the spectral band, but there are large power spectral density components at the low-frequency end of the spectrum which again can lead to baseline wander problems.

3B4B has a spectrum that is fairly evenly distributed throughout the spectral band. It also has good timing content. It is, however, a fairly complex code to implement but, because this class of code can have high efficiency it does have application in long-distance transmission.

4.11 Receiver clock synchronization (clock recovery)

With synchronous transmission each frame is transmitted as a continuous stream of binary bits. It is therefore necessary to introduce a method of clock synchronization such that the receiver and transmitter can keep in step. All the line codes looked at so far are guaranteed to provide sufficient transitions in the bit stream to synchronize a separate clock held at the receiver.

Clock recovery is the extraction of timing information from the received signal. This timing information is used to synchronize the receiver's clock to keep it in step with the transmitter's clock. If the received signal contains a spectral component at the signalling frequency, this can be filtered out and used as a locking signal to synchronize the receiver's clock. If the received signal does not contain a component at the signalling frequency, some sort of nonlinear processing may be used to convert another frequency component to the desired frequency. A simple technique that could be used with, say, AMI or HDB3, neither of which contains a component at the signalling frequency, is to differentiate the incoming signal and to follow this by full-wave rectification. The differentiated signal is series of spikes: one for each transition. This is shown in Figure 4.12(a) and (b) for the message signal consisting of repetitive binary 1s. The differentiated signal still does not contain a component at the signalling frequency and needs to be followed by full-wave rectification to achieve it, as shown in Figure 4.12(c). Figure 4.13 shows a simple circuit to carry out the timing extraction .

A problem that often arises with a clock recovery circuit is that there may not be sufficient timing pulses to produce a good clock waveform. If, for instance, there is a long gap between timing pulses then the clock frequency may drift and errors will occur in the received bit stream. A method that gives more stability to the clock recovery circuit is to use a phased-lock loop recovery circuit as shown in Figure 4.14. A feedback loop is used to control the frequency of the oscillator so that it runs at the correct frequency. The reference signal is the stream of timing pulses derived from the data input. The oscillator still drifts between gaps in the incoming stream of pulses but is more stable than a straightforward oscillator.

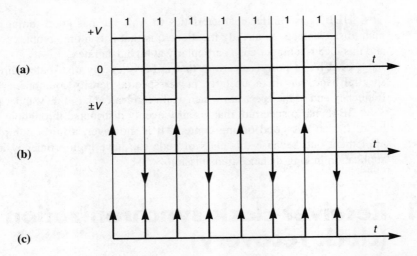

Figure 4.12 Timing extraction for AMI/HDB3 signals: (a) AMI or HDB3 signal (idealized); (b) differentiated signal; (c) full-wave rectified.

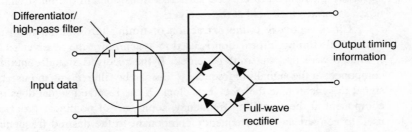

Figure 4.13 Simple circuit for extraction of timing information.

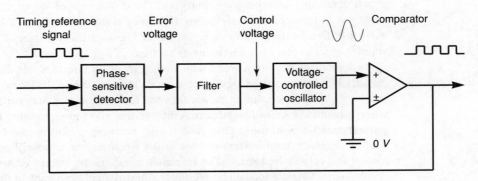

Figure 4.14 Phased-locked loop clock recovery circuit.

4.12 Optical fibre line systems

The codes looked at so far are mostly designed for use in metallic line systems. In the United Kingdom, coaxial cable systems have been almost entirely superseded by optical systems for long-distance digital transmission. The requirements for line codes in optical fibre systems are rather different and therefore need to be considered in their own right.

Optical fibre transmission involves the modulation of a light ray of which the most common form is **On-Off Keying** (OOK). In this system a binary 1 signal is represented by the light being on and a binary 0 by the light being off. All the receiver has to do is to measure the intensity of the received light and check it against a predetermined threshold level. Exceeding the threshold would indicate a binary 1 and not reaching the threshold level would indicate a binary 0.

The required spectral characteristics are also different from those of metallic systems. Optical fibres carry a signal in the form of light either on or off. Hence there is now no concern about baseline wander for the line system although some receivers have features that require the line code to eliminate baseline wander.

Optical fibre systems are wide bandwidth and therefore the need for high-efficiency codes no longer applies as signalling rate is no longer a limiting factor. However, high-effiency codes are of benefit in receiver design where the signalling rate must be kept down.

Line codes in optical fibre systems therefore have the following features:

- binary codes,
- error detection facilities, and
- timing information.

The most common line codes used are nBmB systems, especially 3B4B, although CMI has been used in low-capacity systems.

The transatlantic optical fibre system TAT-8 operates at 280 Mbps and uses a 24B1P line code. This line code breaks the incoming data into blocks of 24 binary bits and adds 1 parity bit to make an odd parity 25-bit block; it could also be described as a 24B25B code. From Section 4.9 the efficiency of this code is given by:

$$\eta = \frac{24}{25} \times 100 = 96\%$$

4.13 Scramblers

Many subsystems in data communications systems, such as equalizers, work better with random bit sequences. Strings of binary 1s or 0s, or periodic sequences may appear in the output of any information source, therefore such sequences need to be coded for transmission if the data transmission systems are likely to have difficulty in conveying them. A device used to randomize the input data is called a **scrambler**. Scramblers are often used in conjunction with some of the line codes described

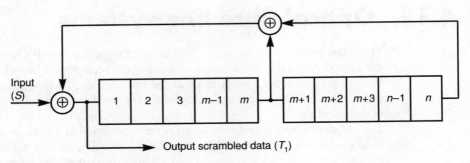

Figure 4.15 Scrambler.

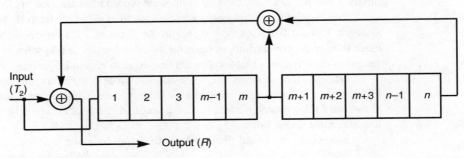

Figure 4.16 Descrambler.

earlier in order to ensure that undesirable sequences are not present in the encoded data. When a scrambler is used at the transmitter a descrambler must be used at the receiver to reconstitute the input data.

A typical scrambler consists of a feedback shift register and the matching descrambler consists of a feed forward shift register. In both the scrambler and the descrambler the outputs of several stages of shift register are added together modulo-2 (exclusive or), and then added to the data stream again in modulo-2 arithmetic. The contents of the shift register are shifted at the bit rate of the system. A block diagram of a typical scrambler is shown in Figure 4.15 and its associated descrambler in Figure 4.16.

The operation of a scrambler may be analysed by allowing the operator D to represent a 1-bit delay. Thus DS represents a sequence S delayed by 1-bit and $D^k S$ represents a sequence S delayed by k bits. Using this method the scrambler and descrambler shown in Figures 4.15 and 4.16 can be analysed as follows.

The output of the scrambler is given by

$$T_1 = S \oplus (T_1 D^m \oplus T_1 D^n)$$
$$= S \oplus F T_1 \quad \text{where } F = D^m + D^n$$
$$= \frac{S}{1 \oplus F}$$

The output of the descrambler is given by:

$$R = T_2 \oplus (T_2 D^m \oplus T_2 D^n)$$
$$= T_2 \oplus F T_2$$
$$= T_2(1 \oplus F)$$

In the absence of errors $T_2 = T_1$ and therefore the unscrambled output is given by:

$$R = \frac{S}{1 \oplus F} \times (1 \oplus F)$$
$$= S$$

Thus the output is an exact duplicate of the input.

ITU-T specifies a scrambler by use of a **tap polynomial** which gives the position of the feedback taps. For example, the scrambler of Figure 4.15 has feedback taken from stages m and n, so it may be described by listing its feedback taps as (m, n). Its tap polynomial is:

$$1 + D^m + D^n$$

Then from the analysis above it can be seen that

$$T_1 = S \text{ (divided by the tap polynomial)}$$

and

$$R = T_2 \text{ (multiplied by the polynomial)}$$

Note that the selection of n for a given m is non-trival and is outside the scope of this book, see Peterson (1961).

EXAMPLE 4.5

Sketch the block diagrams for a scrambler and descrambler with a tap polynomial of

$$1 + D + D^2$$

Figure 4.17 shows a suitable scrambler and descrambler.

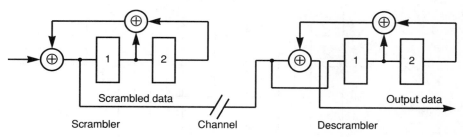

Figure 4.17 Two-stage scrambler and descrambler.

In general, an appropriately designed scrambler of length n will respond to an all zeros input with an output that is a repeating sequence of length $2^n - 1$, provided that it is not latched. A scrambler of this type is called a **maximal-length scrambler**.

A situation can occur in a scrambler where if all the storage elements contain zero and the input is all zeros, then the output will be all zeros. In this (generally undesirable) state the scrambler is said to be latched, and the effect is described as **scrambler lock-up**. It is possible to prevent this condition by using additional circuitry, but an all zeros input is still possible. More complex scrambler configurations ensure that the input conditions that cause lock-up are less likely to occur.

Exercises

4.1   Calculate the radix, redundancy and efficiency of the HDB3 line code.

4.2   (a)   Decode the HDB3 waveform shown in Figure 4.18.

   (b)   Sketch a possible HDB3 waveform of the following binary sequence:
          10000010111100001

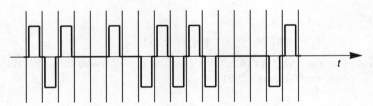

Figure 4.18   HDB3 waveform.

4.3   Sketch a possible CMI waveform of the binary sequence 11110000.

4.4   Decode the CMI waveform shown in Figure 4.19.

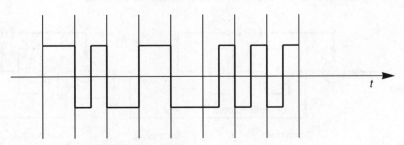

Figure 4.19   CMI waveform.

4.5   Decode the following 3B4B sequence:

$$- - + + \quad + + - - \quad - - + - \quad + - - + \quad + - + + \quad - - + - \quad + - + +$$

4.6   Encode the binary sequence 000001111110000 into 3B4B code:

(a)   assuming the initial running digital sum is +2,

(b)   assuming the initial running digital sum is 0.

4.7   CMI may be interpreted as a 1B2B code by considering the two halves of the encoded symbol as two separate symbols so that a 0 is encoded to $- +$ and a 1 is encoded to $+ +$ or $- -$. Another 1B2B code known as Manchester (or biphase) code has the waveform shown in Figure 4.20.

(a)   Draw up code tables for both CMI and Manchester codes.

(b)   Compare the waveforms of CMI and Manchester encoding of the bit sequence 11110000.

4.8   Sketch a block diagram of the scrambler (and its associated descrambler) described by the tap polynomial $1 + x^3 + x^5$. Given that this is a maximal-length scrambler, what is the (nonlatched) cycle length of the output when the input is all zeros?

4.9   By constructing a table derive the output from a three-stage scrambler with tap polynomial of $1 + x + x^3$ when the input is the repeating sequence 1010:

(a) when the scrambler starts with 1 in the first and third stages and 0 in the second,

(b) when the scrambler starts with 0 in the first stage and 1 in the other two stages.

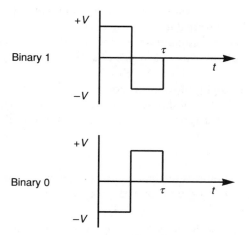

Figure 4.20   Manchester/biphase code rules.

Modems

$$\boxed{5}$$

This chapter discusses the requirements for modems. Various modulation schemes such as FSK and PSK are explained along with reasons for their selection. A brief description of commercially available modems is given along with the appropriate ITU-T standard. The V.24 physical interface is included and the chapter concludes with a look at the new V.34 standard.

5.1 Modem characteristics

The modem's main function is to convert digital signals from a data terminal equipment (DTE) such as a computer into analogue signals suitable for a PSTN telephone line and vice versa. The telephone line was designed for speech communication and has a bandwidth of approximately 300–3400 Hz. This means that the telephone line does not pass the very low frequency signals that may arise from the transmission of a long string of binary 1s or 0s. It is therefore not possible to simply apply different voltages to the telephone line to represent binary 1s and 0s. The binary data must be converted into a form suitable for the telephone line. A circuit which does this is called a **modulator** and the circuit performing the reverse function is called a **demodulator**. Since each side of a data link must normally both send and receive data the combined device is called a **modem**.

Modern modems are highly developed complex pieces of equipment and contain much more than just a modulator and demodulator. Figure 5.1 is a simplified block diagram of a typical modem which contains the following elements:

(1) Transmitter path:

   (a) *Scrambler*   Randomizes the input data in order to prevent long strings of binary 1s or 0s. Chapter 4 provides a description of scramblers and their functions.

   (b) *Encoder*   Encodes the data to provide error protection. Chapter 6 provides a full description of error protection codes.

   (c) *Modulator*   Converts the input digital signal into an analogue signal.

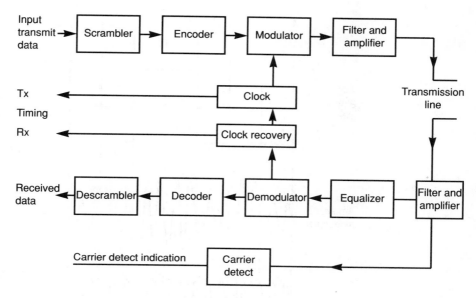

Figure 5.1 Typical modem.

(d) *Filter and amplifier* Ensures that the signal is of the appropriate bandwidth and amplitude for transmission over the telephone lines.

(2) Receiver path:

(a) *Filter and amplifier* Limits the bandwidth to an appropriate level to reduce noise effects and amplifies the input signal to a level suitable for the demodulator.

(b) *Equalizer* Compensates for the limited characteristics of the telephone line such as group delay and amplitude distortion (see Section 5.3).

(c) *Demodulator* Converts the analogue signal back to a digital one.

(d) *Decoder* Checks for errors and removes the redundancy introduced in the coding process.

(e) *Descrambler* Unscrambles the data back to its original form.

(f) *Clock/clock recovery* Ensures that the timing of transmitted and received data is consistent.

5.2 Modulation

There are three types of modulation that are used for converting digital signals to a form suitable for the PSTN. These are amplitude modulation, frequency modulation and phase modulation. Figure 5.2 shows the basic form of each modulation system.

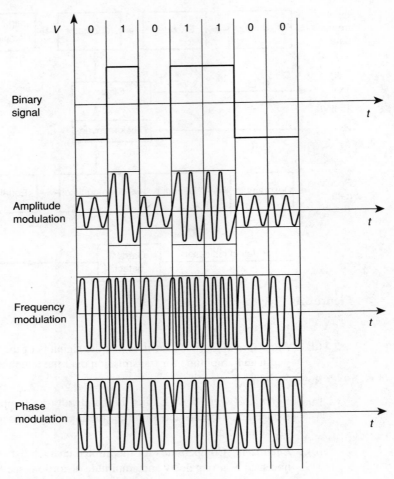

Figure 5.2 Modulation methods.

5.2.1 **Amplitude modulation**

Amplitude modulation is the simplest type of modulation and is implemented in a form known as **Amplitude Shift Keying** (ASK) where a single audio tone is switched between two levels at a rate determined by the input binary data. The signal frequency selected is known as the **carrier frequency** and is chosen to be within the range of frequencies suitable for the PSTN (300–3400 Hz). This type of modulation is useful for illustrative purposes but it is not used in this form in any commercially available modems. This is because it uses amplitude variations to encode and decode binary signals and one of the principal degradations of a tele-phone line is the level of noise it introduces. Noise can dramatically affect the

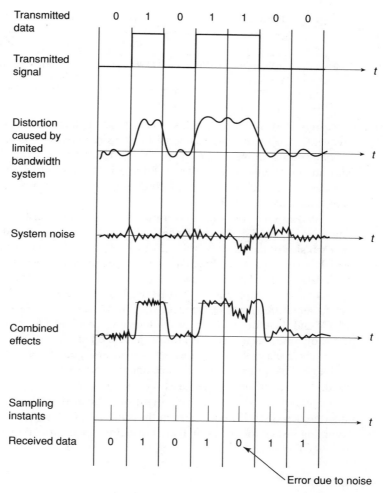

Figure 5.3   Effects of noise on a digital signal.

amplitude of the received signal in such a way that errors are easily introduced, as shown in Figure 5.3. However, amplitude modulation is used in conjunction with phase modulation in more sophisticated modems and is discussed in Section 5.2.4. The bandwidth required for an amplitude modulated signal can be shown to be twice the signalling rate.

5.2.2   Frequency modulation

Since modems need transmit only binary 1s and 0s, data can be represented by switching between two tones of different frequencies. This technique is known as

Table 5.1 ITU-T frequency allocations for a V.21 modem.

	Mark	*Space*	*Carrier*
Originating modem	980 Hz	1180 Hz	1080 Hz
Answering modem	1650 Hz	1850 Hz	1750 Hz

Frequency Shift Keying (FSK). This type of modulation is the method most frequently used with lower data rate modems such as 300 bps or 1200 bps. An FSK modem conforming to the ITU-T V.21 standard (data rate 300 bps) uses the allocation of frequencies shown in Table 5.1. By convention, the modem initiating the call uses the lower carrier frequencies for transmission and the higher frequencies for reception.

A data rate of 300 bps has a maximum fundamental frequency of 150 Hz as shown in Figure 5.4. The frequency spectrum contains sidebands spaced at 150 Hz on each side of the carriers. The V.21 modem has a 200 Hz spacing between carriers which is sufficient to embrace the primary sidebands of each carrier. FSK modems with data rates of 1200 bps have a maximum fundamental component of 600 Hz and have frequency separations of 1000 Hz. For data rates of 9600 bps the maximum fundamental frequency component is 4800 Hz which exceeds the bandwidth of the PSTN and therefore this sort of modulation is not suitable for such speeds. Figure 5.5 shows the spectral diagram for the V.21 modem.

The rectangular-shaped pulses shown in Figure 5.4 are idealized versions; it is not possible to transmit this shape and in practice a more rounded pulse shape is used. A common one is the raised cosine pulse whose shape is that of a positive half-cycle of a cosine wave. This shape of pulse also helps to reduce Inter-Symbol Interference (ISI). ISI occurs when the pulse is spread out in time, such that at the sampling instant for one pulse there is a significant contribution from the succeeding or preceding pulse.

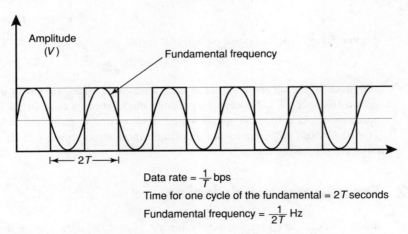

Data rate = $\frac{1}{T}$ bps

Time for one cycle of the fundamental = $2T$ seconds

Fundamental frequency = $\frac{1}{2T}$ Hz

Figure 5.4 Relationship between data rate and fundamental frequency.

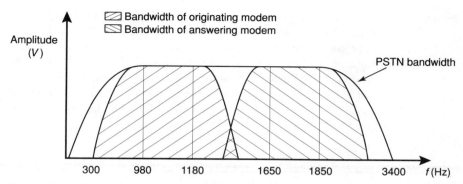

Figure 5.5 Spectral diagram for a V.21 modem.

5.2.3 Phase modulation

With phase modulation the amplitude of the carrier signal is kept constant and the carrier is shifted in phase for each bit of the data stream. This process is known as **Phase Shift Keying** (PSK) . A simple conceptual way to visualize PSK is shown in Figure 5.6 where binary 1 is transmitted direct and binary 0 has a 180° phase shift inserted. This type of modulation is known is also known as **phase-coherent PSK**. In order to decode the incoming signal the receiver needs to compare the phase of a reference carrier signal with the incoming phase of the received signal. To prevent phase errors being introduced the receiver's reference carrier signal needs to be locked in phase with the unmodulated carrier of the transmitter. In practice, such a system can involve complex demodulation circuitry which is expensive to implement. The system is also susceptible to random changes of phase in the received signal. These changes in phase are easily produced in the transmission path and can lead to many errors. To solve this problem, a modified form of PSK is used known as **Differential Phase Shift Keying** (DPSK). In this system the phase is changed relative to the phase of the previous bit transmitted. Hence for a binary 0 the phase is advanced 90° on the phase of the previous bit and for a binary 1 the phase is retarded by 90° on the phase of the previous bit. The demodulator therefore needs to compare only the phase shift between two adjacent bits to determine the binary code.

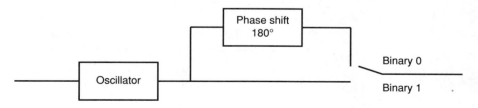

Figure 5.6 Conceptual phase shift keying.

The spectrum of binary digital PSK can be shown to be the same as that of binary ASK. A 180° phase shift of a sinusoidal waveform is equivalent to inverting it, or multiplying it by −1. Hence ASK with signalling amplitudes of +1 and −1 is the same as PSK with signalling levels of 0° and 180°. The spectrum and the bandwidth for binary PSK is therefore equal to that of ASK. It can also be shown that *n*-ary PSK also occupies the same bandwidth as an ASK signal modulated by the same *n*-level baseband signal. So, like ASK, the bandwidth of PSK may be taken to be approximately twice the signalling rate.

EXAMPLE 5.1

For a DPSK system the phase after the last bit was received is +90°. Evaluate the phase shift at the end of the sequence 101101.

Table 5.2 contains the evaluated phase shifts.

Table 5.2 Evaluated phase shifts for Example 5.1.

Binary code	Phase shift	Total phase shift
1	−90°	0°
0	+90°	+90°
1	−90°	0°
1	−90°	−90°
0	+90°	+0°
1	−90°	−90°

DPSK can be extended to encode more states and make more efficient use of bandwidth if each signalling element represents more than one bit. A common encoding technique, known as **Quadrature Phase Shift Keying** (QPSK), uses phase shifts as multiples of 90° as shown in Table 5.3.

As Table 5.3 shows, each signalling state represents two bits rather than one. A differential version is the most common implementation and the phase shifts in Table 5.3 still apply but are now seen as phase changes from the phase of the

Table 5.3 Quadrature phase shift keying.

Phase shift	Binary code
+45°	00
+135°	01
+225°	11
+315°	10

previous transmitted bit. This type of modulation scheme is implemented in the V.26 bis modem and operates at a data rate of 2400 bps in the forward direction and 1200 bps in the reverse direction. An 8-phase (3-bit) version has also been implemented as V.27 and operates at a data rate of 4800 bps.

EXAMPLE 5.2

A Differential Quadrature Phase Shift Keying (DQPSK) system is used to encode the data stream 00 11 01 10. Evaluate the phase shift at the end assuming that the phase shift at the start is 0°.

Table 5.4 lists the evaluated phase shifts.

Table 5.4   Evaluated phase shifts for Example 5.2.

Binary code	Phase shift	Total phase shift
00	+45°	+45°
11	+225°	+270°
01	+135°	+45°
10	+315°	0°

5.2.4 Quadrature amplitude modulation

A technique used in many modern modems is **Quadrature Amplitude Modulation** (QAM), which is a combination of amplitude modulation and phase modulation. It arose out of the realization that DPSK modulation could be represented by two separate baseband signals, with a 90° phase shift, being transmitted on a single carrier with the same frequency. This would be equivalent to simultaneously transmitting information, independently modulated by the in-phase and quadrature components of the same carrier. The incoming digits are mapped into two baseband signals $s_1(t)$ and $s_2(t)$. The baseband signal $s_1(t)$ is multiplied by the in-phase channel $\cos\omega_c t$ and $s_2(t)$ by the quadrature channel $\sin\omega_c t$, the sum of these two products forming the transmitted QAM signal as shown in Figure 5.7.

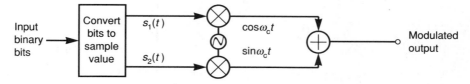

Figure 5.7   QAM Modulation.

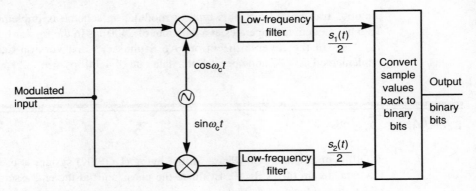

Figure 5.8 QAM Demodulation.

The demodulator reconstitutes the original signal by separately multiplying the received signal by $\cos\omega_c t$ and $\sin\omega_c t$. The first multiplication (after filtering out the high-frequency components) produces $s_1(t)/2$ and the second produces $s_2(t)/2$ as shown in Figure 5.8.

In QAM systems it is important to decide the mapping from the input binary bits to the sample values $s_1(t)$ and $s_2(t)$. A simple way of doing the mapping, for a 2-bit signal, is to map the first bit into a sample s_1, letting binary 1 be +1 V and binary 0 be −1 V and similarly to map the second bit into sample s_2. This is shown in Figure 5.9, which is called a constellation diagram. Then, similarly, for any given integer k of binary bits you can map into two amplitudes s_1 and s_2. A constellation diagram for a 16-point V.29 modem is shown in Figure 5.10. To reduce the likelihood of noise causing one constellation point to be mistaken for another, it is desirable to place the adjacent signal points as far away from each other as possible, subject to constraints on signal power.

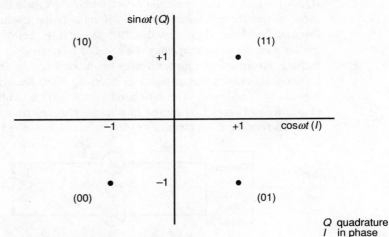

Figure 5.9 Four-point QAM constellation.

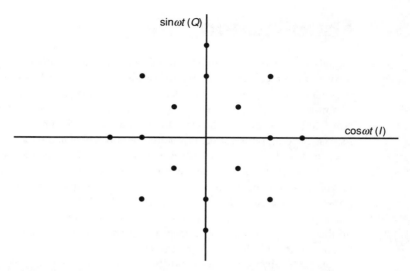

Figure 5.10   Signal constellation for a 16-point V.29 modem at 9600 bps.

5.2.5 Trellis code modulation

For data rates in excess of 9600 bps it is difficult in a QAM system to place the constellation points sufficiently far away from each other to prevent impairments from causing errors. To overcome this problem a new generation of modems based on **Trellis Code Modulation** (TCM) has been developed. In a QAM modem an impairment causes the received signal point to be displaced from its correct location in the constellation. The receiver then selects the signal point in the constellation that is the closest to what has been received. Hence when the impairments are large enough to cause the received point to be closer to a signal point that is different to what was transmitted an error occurs. TCM has an encoder that adds a redundant bit to each symbol in order to minimize these errors.

    For the V.32 bis modem the transmitter breaks the incoming data into 6-bit symbols. Two of the six bits are encoded with a binary **convolutional** encoder. The convolutional encoder adds a code bit to the two input bits forming three encoded bits for each symbol. As a result of this encoding the 6 input bits are converted to 7 encoded bits and are mapped into a 128-point signal constellation.

    The redundancy introduced by the encoder means that only certain sequences of signal points are valid. Thus, if an impairment causes a signal point to be shifted, the receiver compares only the observed point with all valid points and selects the valid point closest to the observed signal. Because of this technique a TCM modem is only half as susceptible to noise power as a conventional QAM modem. Its error rate can be reduced by as much as three orders of magnitude.

5.3 Equalization

Signals cannot be transmitted through any channel without being distorted before arrival at the receiver. When the distortion becomes significant it may no longer be possible to replicate the original transmitted signal at the receiver. The two most common types of degradation are amplitude distortion and phase distortion. Amplitude distortion implies that the attenuation of the channel is not constant over its pass band and this causes problems for both voice and data communications.

Phase distortion has little effect on the intelligibility of speech so any degradation in the phase characteristics of the received signal is of little importance in voice communications. However, for data communication the lack of a linear phase response can lead to an increase in the BER of the channel. The lack of a linear phase response in the channel results in the frequency components that make up the data signal being displaced with respect to each other. This phenomenon is known as **group delay**, where the amount by which a signal of one frequency travels faster than a signal of a different frequency is measured. Group delay in a communications channel may cause a broadening of the binary data pulses which can cause intersymbol interference.

To compensate for the telephone line degradations a circuit known as an **equalizer** is inserted into the modem. The equalizer's function is to cancel out the imperfections of the amplitude and phase characteristic of the channel. An ideal equalizer has a frequency response such that the combination of the channel and equalizer has an approximately constant attenuation and a linear phase response over the frequencies of interest. To achieve this situation the modem needs to know in advance what the frequency response of the channel is. Unfortunately, each time the modem responds to a call the channel is likely to be a different physical path with a different frequency response, so the ideal solution is for an equalizer that can adapt to each different physical path.

For lower data rates, where wider pulses are used and lower frequency components exist, distortion is not too great and it is sufficient to use a **fixed compromise equalizer**. These equalizers are designed to compensate for a channel's typical frequency response and are often located at the transmitter.

At higher data rates an **adaptive equalizer** is used. This type of equalizer can adapt the equalizer characteristic to optimize the signal received by the receiver for a given channel.

The equalizer takes the form of a microprocessor-controlled digital filter which may be adjusted to give optimum performance. The equalizer may be adjusted by one of two methods or by a combination of both.

- *Method 1*   Upon commencement of transmission a short training signal with a predetermined sequence is transmitted. Since the receiver knows what the sequence should be, the equalizer can be adjusted to give the best version of the sequence output. The best version is the one that gives minimum errors.

- *Method 2*   If the frequency response of the channel remains constant method 1 is all that is needed. Unfortunately, frequency response can change for a number of reasons, such as temperature changes. These variations, although small compared to the differences between different physical paths, can still introduce significant errors. A microprocessor-controlled digital filter can track the changes due to the data transmission and can adjust itself for optimum performance. Although in this case it does not know what the transmitted sequence is, it does know that it needs clear decision levels.

5.4 Control

To control the operation of the modem, standard specifications have been laid down by such organizations as ITU-T and the Electronics Industries Association (EIA). The best-known standard is the ITU-T V.24/V.28 interface which is similar to the EIA 232D interface standard more commonly known as RS-232C . Figure 5.11 shows a typical modem interconnection scheme.

V.28 specifies the mechanical and electrical characteristics while V.24 gives the functional and procedural characteristics

The mechanical specification gives details of the physical connection of the DTE and DCE. The specification calls for a 25-pin connector and lays down which signal is connected to which pin. It also specifies that the DTE has a male connector and that the DCE has a female connector. The standard connector used is a 25-pin D-type as shown in Figure 5.12. The arrangement and function of each lead is shown in Table 5.5.

An important point to remember when dealing with V.24 is that all the signal names are viewed from the DTE. Thus the DTE transmits data on pin 2 and receives data on pin 3. For the DCE the pin connections are reversed; the DCE transmits data on pin 3 and receives data on pin 2.

The function of each line is as follows:

- Pins 1 and 7   Pin 7 is the signal ground and is the common reference for all signals. *The interface will not work unless this pin is connected.* It must be connected at both the DTE and the DCE ends. Pin 1 is connected to the chassis of the equipment and is intended to be connected to one end of a shield if shielded cable is used. A shielded cable may be used to minimize interference in high-noise environments. The use of two separate grounds can cause earth loops to occur and it is better to leave pin 1 unconnected in most systems.

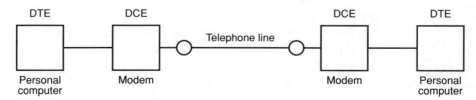

Figure 5.11 Modem interconnection scheme.

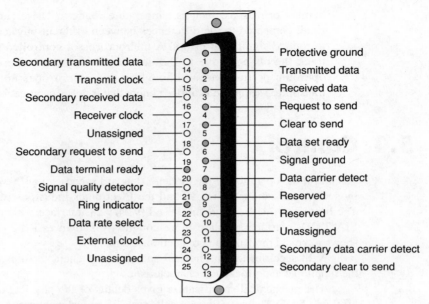

Secondary transmitted data — 14
Transmit clock — 15
Secondary received data — 16
Receiver clock — 17
Unassigned — 18
Secondary request to send — 19
Data terminal ready — 20
Signal quality detector — 21
Ring indicator — 22
Data rate select — 23
External clock — 24
Unassigned — 25

1 — Protective ground
2 — Transmitted data
3 — Received data
4 — Request to send
5 — Clear to send
6 — Data set ready
7 — Signal ground
8 — Data carrier detect
9 — Reserved
10 — Reserved
11 — Unassigned
12 — Secondary data carrier detect
13 — Secondary clear to send

Figure 5.12 Typical D-type V.24/V.28 connector.

- Pins 2 and 3   The DTE transmits data on pin 2 and receives on pin 3.

- Pins 4 and 5   The DTE asserts an RTS signal to the DCE; when the DCE has turned its transmission carrier on and is ready to accept data it sends CTS back to the DTE.

- Pins 6, 20 and 22   DSR indicates that the DCE is switched on and is not in a test mode. A dial-up DCE is in the 'off-hook' mode and any timing functions required are completed. The DTR may be used in response to a ring indicator, RI, to tell the DCE to answer the incoming call. The RI is the way the DCE tells the DTE that it is receiving an incoming call.

- Pin 8   This indicates that the DCE is receiving a signal of acceptable quality.

The above pin connections are by far the most often used in V.24 applications and little use is made of the following pin connections:

- Pins 15, 17 ,21 and 24   These connections are used by synchronous DCEs to control bit timing.

- Pin 23   This is used in applications where a DCE can operate at two different data rates and the signal on pin 23 controls whether high or low speed is used.

- Pins 12, 13, 14, 16 and 19   Some DCEs are equipped with both primary and secondary channels. All these pins are associated with the secondary channel operation. In these applications the primary channel usually has the higher data rate, and the secondary channel transmits in the reverse direction at a much lower data rate (for example, 75 bps).

Table 5.5 V.24/V.28 pin designations.

Pin no.	Signal description	Abbreviation	From DCE	To DCE
1	Chassis ground	GND		Yes
2	Transmitted data	TD		Yes
3	Received data	RD	Yes	
4	Request to send	RTS		Yes
5	Clear to send	CTS	Yes	
6	Data set ready	DSR	Yes	
7	Signal ground	SG	Yes	Yes
8	Data carrier detect	DCD	Yes	
9	Test			
10	Test			
11	Unused			
12	Secondary channel received line signal detector	SEC RLSD	Yes	
13	Secondary channel received data	SEC RD	Yes	
14	Secondary channel transmitted data	SEC TD		Yes
15	Transmitter timing	TX TIMING	Yes	
16	Secondary channel received data	SEC RD	Yes	
17	Receiver timing	RX TIMING		
18	Unused			
19	Secondary channel request to send	SEC CTS		Yes
20	Data terminal ready	DTR		Yes
21	Signal quality detector	SQ	Yes	
22	Ring indicator	RI	Yes	
23	Rate		Yes	Yes
24	Transmitter timing		Yes	
25	Test			

The electrical characteristics specify the signalling between DTE and DCE. Digital signalling is used with the following conventions:

- The transmitter generates a voltage of between +5 V and +25 V for a binary 0 and a voltage of between −5 V and −25 V for a binary 1.

- The receiver circuit recognizes voltages of above +3 V as binary 0 and voltages below −3 V as binary 1.

- The data rate should be less than 20 kbps and the distance between a DTE and a DCE should be less than 15 m.

- In practice the actual voltage levels used are determined by the supply voltages and ±12 V or even ±15 V are not uncommon.

The most common implementation of V.24 is as a primary channel asynchronous data interface. A very simple implementation of this without any handshaking is just to connect the data and signal ground lines as shown in Figure 5.13.

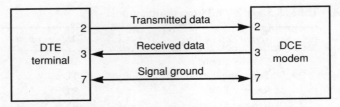

Figure 5.13 Minimum V.24 interface connections.

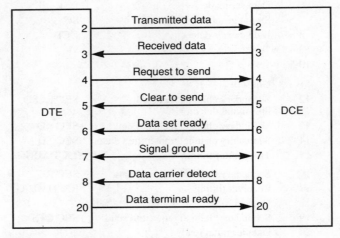

Figure 5.14 Typical asynchronous system connections.

However, many applications require more control functions to enable greater control over the flow of data and Figure 5.14 shows the typical pin connections required in this case.

The V.24 interface has achieved such universal use that it is often found in applications where there is no obvious DCE. For example, many manufacturers specify a V.24 connector for a computer-to-printer cable. In a sense both these devices are DTEs and there is no obvious DCE. A question that has to be resolved is whether pin 2 or pin 3 is used for data transfer. Fortunately, for this application, most of the signals defined in V.24 are unnecessary and the manufacturer's special cable will have many pins unconnected and some of the pin connections will be jumped and interchanged in order to adapt the interface at both ends.

5.5 V-standards

Most of the present computer communications standards have been developed through organizations such as the ITU, which sets communications recommendations for the world. Its specifications are characterized by the letter 'V' for the

PSTN and 'X' for the PDN. 'V' simply means standards, with the number following indicating the particular standard. The term 'bis' is used to indicate a second version of a particular standard and the term 'terbo' a third version. Table 5.6 shows a summary of the V-standards as they apply to modems.

In the mid-1980s work began on a new modem standard provisionally called V.fast but now implemented as V.34. The initial aim was to achieve data rates of 19 200 bps but this was later raised to 28 800 bps. These high data rates were made feasible by the general upgrading of the PSTN and by modern technologies beyond those used in V.32.

V.34 has built upon the rapid advances that have been made in Digital Signal Processing (DSP) in the last few years. V.34 uses advanced features such as precoding, multidimensional trellis coding and new constellation mapping/shaping techniques. To achieve its highest data rate it uses the full voice bandwidth offered and adjusts the signalling bandwidth and data rate accordingly. This means the use of automatic adaptation of the modulation/coding to suit the conditions encountered.

Modem development since the early 1960s has shown a dramatic increase in the data rates achieved. The modems produced in the 1960s had data rates of 300 bps whereas commonly available modems in 1994 had data rates of 14 400 bps and the new V.34 standard has a data rate of 28 800 bps. However, with the development of ISDN and dedicated digital lines it is anticipated that the modem will gradually disappear and direct connections to the digital network will be made. The maximum data rate currently available on narrowband ISDN is 128 kbps. However, modems still have a very healthy market and will be used well into the next century.

Table 5.6 Modem V-standards.

V-series	Data rate bps	Modulation	Equalization
V.21	300	FSK (2 frequencies)	–
V.23	1200/600	FSK (2 frequencies)	–
V.26 bis	2400/1200	4-phase DPSK	fixed compromise
V.27	4800	8-phase DPSK	manually adjusted
V.27 ter	488/2400	8-phase DPSK	automatic adaptive
V.29	up to 9600	16-point constellation QAM	automatic adaptive
V.32	up to 9600	16-point QAM or 32-point (redundant) TCM	automatic adaptive
V.32 bis	up to 14400	up to 128-point TCM	adaptive with echo cancellation
V.34	$n \times 2400$ up to 28 800	precoding with multipoint 4-dimensional TCM	to include adaptive bandwidth feature

Exercises

5.1 A phase modulation system, where a binary 1 is transmitted direct and a binary 0 has a 180° phase shift inserted, has the following binary sequence applied to it:

101100101

Sketch the modulated wave form clearly showing the phase changes.

5.2 A DPSK system advances the phase shift by 90° for a binary 0 and by 270° for a binary 1. Assuming that the previous bit had a phase shift of −90°, evaluate the phase shift for the last bit in the data stream 101100101.

5.3 Estimate the bandwidth required for a data rate of 1200 bps for the following modulation schemes:

(a) ASK
(b) PSK
(c) DPSK

Error control

6

In Chapter 3 we saw how information theory and source coding allow us to optimize the information content of transmitted data and thus increase the efficiency of the transmission. However, once the data is dispatched over the transmission medium the characteristics of the medium normally conspire to alter the transmitted data in various ways so that the signals received at the remote end of a link differ from the transmitted signals. These adverse characteristics of a medium are known as **transmission impairments** and they often reduce transmission efficiency. In the case of binary data they may lead to errors, in that binary 0s are transformed into binary 1s and vice versa. To overcome the effects of such impairments it is necessary to introduce some form of error control. The first step in any form of error control is to detect whether any errors are present in the received data; a process which will be explored in some detail in Section 6.2. Having detected the presence of errors there are two strategies commonly used to correct them: either further computations are carried out at the receiver to correct the errors, a process known as **forward error control**; or a message is returned to the transmitter indicating that errors have occurred and requesting a retransmission of the data, which is known as **feedback error control**. Error control is a function of the data link layer of the ISO model for OSI.

6.1 Transmission impairments

A signal which is received at the remote end of a link will differ from the transmitted signal as a result of transmission impairments introduced by the transmission medium. The three main impairments are **noise**, **distortion** and **attenuation**.

6.1.1 Noise

The main factor which constrains the operation of any communications system is the presence in a communication channel of random, unwanted signals, known as noise. One form of noise, **thermal noise**, is present in all electronic devices and metallic transmission media, as a result of the thermal agitation of electrons within

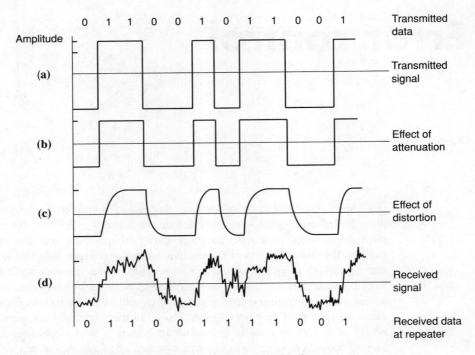

Figure 6.1   Effects of attenuation, distortion and noise.

the material. It can never be eliminated and places an upper limit on the performance of a communications system. Probably the most troublesome form of noise in a data link is **impulse noise**. This consists of electrical impulses in a communications channel which are picked up from external sources such as lightning or noisy electromechanical equipment. The problem with impulse noise is that, although normally infrequent, it can have a fairly high amplitude and can be of a relatively long duration thus corrupting substantial numbers of bits. Although it is, strictly speaking, an example of **interference**, **crosstalk** produces similar problems to noise. It arises from unwanted electrical coupling between adjacent communications channels. Its most obvious manifestation is when some other conversation can be heard in the background of a telephone call. The effect of noise on data signals is illustrated in Figure 6.1(d).

6.1.2 Distortion

Signals transmitted along a transmission line suffer various kinds of distortion. In an electrical circuit the inherent capacitance produces the familiar changes of shape in a signal illustrated in Figure 6.1(c). A further type of distortion occurs in all guided transmission circuits because different signal frequencies travel at different velocities within a waveguide and hence arrive at a receiver at slightly different times. This effect is known as **delay distortion** in electrical circuits and **dispersion** in optical circuits. If distortion is severe enough, received bits can spread into adjacent bit positions, an effect known as **intersymbol interference**.

6.1.3 Attenuation

As a signal is transmitted over a transmission medium its amplitude decreases, a phenomenon known as attenuation. The effect of attenuation, noise and distortion on data signals is shown in Figure 6.1(d). If an electrical signal is transmitted over a distance of more than a few kilometres then the attenuation and distortion are such that noise might cause erroneous reception of data. Thus, metallic data circuits have repeaters spaced at regular intervals which take attenuated (and distorted) signals, such as those in Figure 6.1(d), and repeat them as perfect signals, as in Figure 6.1(a). Note, however, that in this instance, the repeater will retransmit an erroneous bit in position 7 as a result of the transmission impairments. A major advantage of optical circuits is that attenuation is dramatically less than in other media, resulting in links of 100 km or more without repeaters. Attenuation in electrical circuits is both optical and a function of frequency, which introduces further distortion in a received signal as some frequency components within a signal are attenuated more than others. This problem can be overcome by the use of equalizers (see Chapter 5) which equalize the amount of attenuation across the range of frequencies used.

6.2 Forward error control

The need to detect and correct errors was mentioned in Section 2.1.7, but not the means by which the detection process can be carried out. Error detection (and correction) is achieved using **channel coding**. Channel coding is the process of coding data prior to transmission over a communications channel so that if errors do occur during transmission it is possible to detect and possibly even to correct those errors once the data has been received. In order to achieve this error detection/correction some bit patterns need to be identified as error free at the receiver, whereas other bit patterns will be identified as erroneous. To increase the number of identifiable bit patterns at the receiver above the bare minimum required to represent the data, additional bits, known as **redundant bits**, are added to the data or information bits prior to transmission. Various different types of code are available for use in channel coding but the most commonly used are called **linear block codes**.

6.3 Linear block codes

These constitute the simplest and most commonly used type of channel code. Data is transmitted as a fixed-length block. Prior to transmission the data is treated as a binary number and some form of linear mathematical process is carried out on a group of information bits so as to generate additional redundant bits which are known as **check bits**.

The check bits are transmitted along with the information bits, normally at the end of the block. At the receiver, a similar mathematical process is used to determine whether there are errors or not. Typical mathematical processes used are addition and division. A study of these codes inevitably requires some knowledge of mathematics but although the theory underlying the codes is extremely complex, the following treatment has been kept fairly straightforward without being too simplistic.

6.3.1 Hamming codes

This is an important group of early error-correcting codes pioneered by R.W. Hamming in the 1950s. They involve the production of check bits by adding together different groups of information bits. The type of addition used is modulo-2 and is equivalent to normal binary addition without any carries. The best way to see how the check bits are obtained is to consider a particular code as an example. We shall consider a Hamming (7,4) code, in which three check bits (c_1, c_2, and c_3) are combined with four information bits (k_1, k_2, k_3 and k_4) to produce a block of data of length $n = 7$. This block of data is known as a code word. A block of data of length 7 is too short to be appropriate for a practical data communications system, but the mathematics involved in longer blocks would become tedious. Three check equations are used to obtain the three check bits of this Hamming (7,4) code as follows:

$$c_1 = k_1 \oplus k_2 \oplus k_4$$
$$c_2 = k_1 \oplus k_3 \oplus k_4$$
$$c_3 = k_2 \oplus k_3 \oplus k_4$$

where $\oplus$ represents modulo-2 addition. The rules of modulo-2 addition are:

$$0 \oplus 0 = 0$$
$$0 \oplus 1 = 1$$
$$1 \oplus 1 = 0 \text{ (no carry)}$$

If we choose the information bits 1010 as an example then $k_1 = 1$, $k_2 = 0$, $k_3 = 1$ and $k_4 = 0$ and the check bits obtained from the three check equations above as follows:

$$c_1 = k_1 \oplus k_2 \oplus k_4 = 1 \oplus 0 \oplus 0 = 1$$
$$c_2 = k_1 \oplus k_3 \oplus k_4 = 1 \oplus 1 \oplus 0 = 0$$
$$c_3 = k_2 \oplus k_3 \oplus k_4 = 0 \oplus 1 \oplus 0 = 1$$

The code word is obtained by adding the check bits to the end of the information bits and therefore the data 1010101 will be transmitted (information bits first).

A complete set of code words can be obtained in a similar way:

Code word no.	k_1 k_2 k_3 k_4 c_1 c_2 c_3	Code word no.	k_1 k_2 k_3 k_4 c_1 c_2 c_3
0	0 0 0 0 0 0 0	8	1 0 0 0 1 1 0
1	0 0 0 1 1 1 1	9	1 0 0 1 0 0 1
2	0 0 1 0 0 1 1	10	1 0 1 0 1 0 1
3	0 0 1 1 1 0 0	11	1 0 1 1 0 1 0
4	0 1 0 0 1 0 1	12	1 1 0 0 0 1 1
5	0 1 0 1 0 1 0	13	1 1 0 1 1 0 0
6	0 1 1 0 1 1 0	14	1 1 1 0 0 0 0
7	0 1 1 1 0 0 1	15	1 1 1 1 1 1 1

An error that occurs in a transmitted code word can be detected only if the error changes the code word into some other bit pattern that does not appear in the code. This means that the code words transmitted over a channel must differ from each other in at least two bit positions. If two code words differ in only one position and an error occurs in that position then one code word will be changed into another code word and there will be no way of knowing that an error has occurred. Inspection of the set of code words of the Hamming (7,4) code reveals that they all differ from each other in at least three places. Taking code words 3 and 8 as an example, we have:

Code word 3   0 0 1 1 1 0 0
Code word 8   1 0 0 0 1 1 0

These two code words differ in positions 1, 3, 4 and 6 (counting from the left). The number of positions by which any two code words in a code differ is known as the **Hamming distance** or just the distance, so that the distance between these two words is four. Since all linear block codes contain the all-zeros code word, then an easy way to find the **minimum distance** of a code is to compare a nonzero code word which has the minimum number of 1s with the all-zeros code word. Thus the minimum distance of a code is equal to the smallest number of 1s in any nonzero code word, which in the case of this Hamming (7,4) code is three. If the code words of a code differ in three or more positions then error correction is possible since an erroneous bit pattern will be 'closer' to one code word than another (this assumes that one error is more likely than two, two more likely than three, and so on). If we take code words 8 and 10 as an example, we have:

Code word 8   1 0 0 0 1 1 0
Code word 10   1 0 1 0 1 0 1

The distance between these two code words is three. If code word 8 is transmitted and an error occurs in bit 3 then the received data will be:

1 0 1 0 1 1 0

This is not one of the 16 Hamming (7,4) code words since an error has occurred. Furthermore, the most likely code word to have been transmitted is code word 8 since this is the nearest to the received bit pattern. Thus, it should also be possible to correct the received data by making the assumption that the transmitted code word was number 8. If, however, a second error occurs in bit 7 then the received bit pattern will be:

1 0 1 0 1 1 1

It should still be possible to detect that an error has occurred since this is not one of the 16 code words. However, it is no longer possible to correct the errors since the received bit pattern has changed in two places and is no longer closer to code word 8 than any other (it is, in fact, now closer to code word 10). Thus, this Hamming (7,4) code is able to detect two errors but correct only one error. In general, if the minimum distance of a code is d, then $d - 1$ errors can normally be detected using a linear block code and mod$(d - 1)/2$ can be corrected.

A feature of all linear block codes which arises out of the mathematical rules used to determine the check bits is that all the code words are related by these rules. Received data which contains errors does not have this mathematical property, and it is this fact that is used to carry out the detection and correction processes at the receiver. If we take the example of the Hamming (7,4) code then the encoding process will restrict the number of different code words that can be transmitted to the 16 listed above. As a result of errors that may occur in the transmission process, the data arriving at a receiver in 7-bit blocks can have any one of $2^7 = 128$ different 7-bit patterns. This allows the receiver to detect whether errors have occurred since it is aware of the rules used in the encoding process and can apply them to see whether the received data is one of the 16 'legal' code words or not.

Furthermore, it is the case that all Hamming codes (indeed, all linear block codes) possess the mathematical property that if we add any two code words together (modulo-2 addition) then the resulting sum is also a code word. For example, if we add code words 1 and 2 from the earlier list we obtain:

0 0 0 1 1 1 1
⊕ 0 0 1 0 0 1 1
0 0 1 1 1 0 0   which is code word 3

This allows us to represent a whole code by means of a small 'subset' of code words, since further code words can simply be obtained by modulo-2 addition. In the case of the Hamming (7,4) code this is not important, since there are only 16 code words. However, with longer block lengths the number of code words becomes unmanageable. For example, a short block of 32 bits involves $2^{32} = 4\,294\,967\,296$ different code words. The subset of code words is often expressed as a matrix known as a **generator matrix**, G. The code words chosen are normally powers of 2, that is code words 1, 2, 4, 8, ... A suitable generator matrix for the Hamming (7,4) code consists of the following four code words:

$$G = \begin{bmatrix} 1 & 0 & 0 & 0 & 1 & 1 & 0 \\ 0 & 1 & 0 & 0 & 1 & 0 & 1 \\ 0 & 0 & 1 & 0 & 0 & 1 & 1 \\ 0 & 0 & 0 & 1 & 1 & 1 & 1 \end{bmatrix}$$

The matrix has 4 rows and 7 columns, that is, it has dimensions 4×7, ($k \times n$). The whole code can be generated from this matrix just by adding together rows, and it is for this reason that it is called a generator matrix. A further reason for the generator matrix being so named is that it can be used to generate code words directly from the information bits without using the check equations. This is achieved by multiplying the information bits by the generator matrix using matrix multiplication, as Example 6.1 shows.

EXAMPLE 6.1

Information consisting of the bits 1010 are to be encoded using the Hamming (7,4) code. Use the generator matrix to obtain the code word to be transmitted.

The code word is obtained by multiplying the four information bits (expressed as a row vector) by the generator matrix as follows:

$$[\,1 \ 0 \ 1 \ 0\,] \times \begin{bmatrix} 1 & 0 & 0 & 0 & 1 & 1 & 0 \\ 0 & 1 & 0 & 0 & 1 & 0 & 1 \\ 0 & 0 & 1 & 0 & 0 & 1 & 1 \\ 0 & 0 & 0 & 1 & 1 & 1 & 1 \end{bmatrix}$$

The multiplication is achieved by multiplying each column of the generator matrix in turn by the row vector as follows:

$[(1 \times 1 + 0 \times 0 + 1 \times 0 + 0 \times 0),(1 \times 0 + 0 \times 1 + 1 \times 0 + 0 \times 0),(1 \times 0 + 0 \times 0 + 1 \times 1 + 0 \times 0),(1 \times 0 + 0 \times 0 + 1 \times 0 + 0 \times 1),(1 \times 1 + 0 \times 1 + 1 \times 0 + 0 \times 1),(1 \times 1 + 0 \times 0 + 1 \times 1 + 0 \times 1),(1 \times 0 + 0 \times 1 + 1 \times 1 + 0 \times 1)] = 1\ 0\ 1\ 0\ 1\ 0\ 1$

Note that this process is, in fact, the same as adding together the first and third rows of the matrix. There are thus three different ways in which encoding may be carried out in a Hamming code:

(1)   Use the check equations to obtain the check bits and then add the check bits to the end of the information bits.

(2)   Add together appropriate rows from the generator matrix.

(3)   Multiply the information bits by the generator matrix.

It is also possible to express the three check equations of the Hamming (7,4) code in the form of a matrix known as the check matrix H, as follows:

Check equations

$$c_1 = k_1 \oplus k_2 \oplus k_4$$
$$c_2 = k_1 \oplus k_3 \oplus k_4$$
$$c_3 = k_2 \oplus k_3 \oplus k_4$$

Check matrix

$$\begin{bmatrix} 1 & 1 & 0 & 1 & 1 & 0 & 0 \\ 1 & 0 & 1 & 1 & 0 & 1 & 0 \\ 0 & 1 & 1 & 1 & 0 & 0 & 1 \end{bmatrix} = H$$

$$k_1 \ k_2 \ k_3 \ k_4 \ c_1 \ c_2 \ c_3$$

The check matrix is obtained by having each row of the matrix correspond to one of the check equations in that if a particular bit is present in an equation, then that bit is marked by a 1 in the matrix. This results in a matrix with dimensions 3×7 ($c \times n$). If we now compare the two types of matrix we note that the generator matrix has an identity matrix consisting of a diagonal or **echelon** of 1s to its left and the check matrix has this echelon to its right. When a generator or check matrix conforms to this pattern, it is in **standard echelon form**. In the case of single error-correcting codes such as the Hamming codes it usually makes calculations easier if the matrices are in this form. A further point to note is that if the echelons are removed from the two matrices, then what remains is the transpose of each other. In the case of the Hamming (7,4) code:

$$\begin{bmatrix} 1 & 1 & 0 & 1 \\ 1 & 0 & 1 & 1 \\ 0 & 1 & 1 & 1 \end{bmatrix} \quad \text{is the transpose of} \quad \begin{bmatrix} 1 & 1 & 0 \\ 1 & 0 & 1 \\ 0 & 1 & 1 \\ 1 & 1 & 1 \end{bmatrix}$$

EXAMPLE 6.2

The generator matrix for a Hamming (15,11) code is as follows:

$$G = \begin{bmatrix} 1 & 0 & 0 & 0 & 0 & 0 & 0 & 0 & 0 & 0 & 0 & 1 & 1 & 0 & 0 \\ 0 & 1 & 0 & 0 & 0 & 0 & 0 & 0 & 0 & 0 & 0 & 0 & 1 & 1 & 0 \\ 0 & 0 & 1 & 0 & 0 & 0 & 0 & 0 & 0 & 0 & 0 & 0 & 0 & 1 & 1 \\ 0 & 0 & 0 & 1 & 0 & 0 & 0 & 0 & 0 & 0 & 1 & 1 & 0 & 1 \\ 0 & 0 & 0 & 0 & 1 & 0 & 0 & 0 & 0 & 0 & 1 & 0 & 1 & 0 \\ 0 & 0 & 0 & 0 & 0 & 1 & 0 & 0 & 0 & 0 & 0 & 1 & 0 & 1 \\ 0 & 0 & 0 & 0 & 0 & 0 & 1 & 0 & 0 & 0 & 1 & 1 & 1 & 0 \\ 0 & 0 & 0 & 0 & 0 & 0 & 0 & 1 & 0 & 0 & 0 & 1 & 1 & 1 \\ 0 & 0 & 0 & 0 & 0 & 0 & 0 & 0 & 1 & 0 & 0 & 1 & 1 & 1 & 1 \\ 0 & 0 & 0 & 0 & 0 & 0 & 0 & 0 & 0 & 1 & 0 & 1 & 0 & 1 & 1 \\ 0 & 0 & 0 & 0 & 0 & 0 & 0 & 0 & 0 & 0 & 1 & 1 & 0 & 0 & 1 \end{bmatrix}$$

Obtain the check matrix of this code.

The code has a block length $n = 15$, consisting of $k = 11$ information bits and $c = 4$ check bits. The generator matrix has dimensions 11×15, and includes a 11×11 identity matrix (an echelon) to the left. The check matrix has dimensions 4×15 and contains a 4×4 identity matrix to its right-hand side. The rest of the check matrix is obtained by removing the identity matrix from the generator matrix and transposing what is left. This results in the last four bits of the first row of the generator matrix becoming the first column of the check matrix and the last four bits of the last row of the generator matrix becoming the 11th column of the check matrix. The check matrix is as follows:

$$H = \begin{bmatrix} 1 & 0 & 0 & 1 & 1 & 0 & 1 & 0 & 1 & 1 & 1 & 1 & 0 & 0 & 0 \\ 1 & 1 & 0 & 1 & 0 & 1 & 1 & 1 & 1 & 0 & 0 & 0 & 1 & 0 & 0 \\ 0 & 1 & 1 & 0 & 1 & 0 & 1 & 1 & 1 & 1 & 0 & 0 & 0 & 1 & 0 \\ 0 & 0 & 1 & 1 & 0 & 1 & 0 & 1 & 1 & 1 & 1 & 0 & 0 & 0 & 1 \end{bmatrix}$$

Having looked in some detail at several ways of encoding information bits, we shall turn our attention to the decoding process which takes place at the remote end of a channel. To determine whether received data is error free or not, it is necessary for all of the check equations to be verified. This can be done by recalculating the check bits from the received data or, alternatively, received data can be checked by using the check matrix. As its name implies, the check matrix can be used to check the received data for errors in a similar way to using the generator matrix to generate a code word. The check matrix, H, is multiplied by the received data expressed as a column vector:

$$\begin{bmatrix} 1 & 1 & 0 & 1 & 1 & 0 & 0 \\ 1 & 0 & 1 & 1 & 0 & 1 & 0 \\ 0 & 1 & 1 & 1 & 0 & 0 & 1 \end{bmatrix} \times \begin{bmatrix} 1 \\ 0 \\ 0 \\ 0 \\ 1 \\ 1 \\ 0 \end{bmatrix}$$

H matrix   Data vector

This time the multiplication is achieved by multiplying each row of the check matrix in turn by the received data vector, as follows:

$$\begin{aligned} (1\times1 + 1\times0 + 0\times0 + 1\times0 + 1\times1 + 0\times1 + 0\times0) \\ (1\times1 + 0\times0 + 1\times0 + 1\times0 + 0\times1 + 1\times1 + 0\times0) \\ (0\times1 + 1\times0 + 1\times0 + 1\times0 + 0\times1 + 0\times1 + 1\times0) \end{aligned} = \begin{bmatrix} 0 \\ 0 \\ 0 \end{bmatrix}$$

If, as is the case here, the received data is error free then the result of this multiplication is zero. This result, which in this case is a 3-bit vector, is known as the **syndrome**. 'Syndrome' is a medical term, and its use in this context seems a bit idiosyncratic. The word 'diagnosis' seems more appropriate as what we are referring to is something that gives us information on the nature of a problem that we wish to cure. (The problem being the presence of an error.) We can now ask the question: what will the syndrome tell us if an error occurs during the data transmission? Suppose that coded data is transmitted as a 7-bit code word $[t]$ and an erroneous 7-bit stream $[r]$ is received. The error that caused the erroneous received data can also be expressed as a vector $[e]$ which contains zeros apart from the error position in which there is a 1.

Then the received data can be thought of as the transmitted code word plus the error vector:

$$[r] = [t] + [e]$$

These vectors can be multiplied by the check matrix H, as follows:

$$[H][r] = [H][t] + [H][e]$$

But $[H][t]$ is the check matrix multiplied by an error-free code word which, by definition, is equal to zero so that:

$$[H][r] = [H][e]$$

Thus $[H][r]$, which is the check matrix multiplied by the received data (that is, the syndrome) will give some indication as to the nature of the error. It is this basic assumption which forms the basis of error correction. This may seem a bit complicated at first sight but Example 6.3 shows that, in practice, it is not too difficult.

EXAMPLE 6.3

Information consisting of four 1s is to be transmitted using the Hamming (7,4) code.

(a) Determine the transmitted code word.

(b) If an error occurs in bit 4 during transmission, determine the syndrome.

(c) Show how the syndrome can be used to correct the error.

(a) To determine the transmitted code word we can use the three check equations to determine the check bits.

$$c_1 = k_1 \oplus k_2 \oplus k_4 = 1 \oplus 1 \oplus 1 = 1$$
$$c_2 = k_1 \oplus k_3 \oplus k_4 = 1 \oplus 1 \oplus 1 = 1$$
$$c_3 = k_2 \oplus k_3 \oplus k_4 = 1 \oplus 1 \oplus 1 = 1$$

The transmitted code word is therefore seven 1s as follows:

$$
\begin{array}{ccccccc}
k_1 & k_2 & k_3 & k_4 & c_1 & c_2 & c_3 \\
1 & 1 & 1 & 1 & 1 & 1 & 1
\end{array}
$$

(b)   If an error occurs in bit 4 then the received data is 1110111. To check whether there is an error in the received data, we multiply by the check matrix:

$$
\begin{bmatrix} 1 & 1 & 0 & 1 & 1 & 0 & 0 \\ 1 & 0 & 1 & 1 & 0 & 1 & 0 \\ 0 & 1 & 1 & 1 & 0 & 0 & 1 \end{bmatrix} \times \begin{bmatrix} 1 \\ 1 \\ 1 \\ 0 \\ 1 \\ 1 \\ 1 \end{bmatrix} = \begin{bmatrix} 1 \\ 1 \\ 1 \end{bmatrix} \text{ Syndrome}
$$

↑
Column 4

(c)   The fact that an error has occurred has caused the syndrome to be non-zero. Furthermore the position of the error can be located by comparing the syndrome with the columns of the check matrix. In this case the syndrome provides us with all we need to know about the error since its numerical value equates with column 4 of the check matrix, thus indicating an error in bit 4.

Thus all of the mathematics that we have used so far proves to be worth while since Hamming codes provide an easily implemented way not only of detecting an error but also of locating its position.

Encoding and decoding circuits

An attractive feature of Hamming codes in the early days of error correction was that they could be easily implemented in hardware circuits, particularly if the block length was fairly short. Since modulo-2 addition is identical to an exclusive-or (EX-OR) function, a number of multiple input EX-OR gates can be used to determine the check bits, as in the circuit for a Hamming (7,4) encoder shown in Figure 6.2.

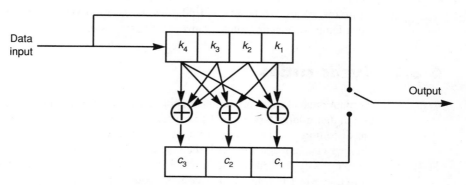

Figure 6.2   Hamming (7,4) encoder.

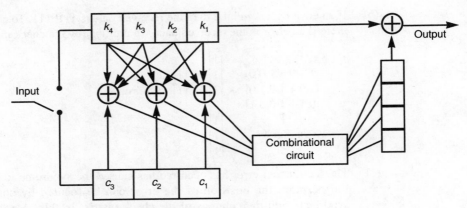

Figure 6.3 Hamming (7,4) decoder.

The information bits are fed into a 4-bit shift register and the check bits are calculated by the EX-OR circuits and are held in a 3-bit shift register. The switch moves up to transmit the information bits and down for the check bits. Figure 6.3 shows a corresponding decoder.

With the receive switch up the information bits are received into a 4-bit shift register and with it down the check bits flow into a 3-bit shift register. The left-hand EX-OR gate works out the modulo-2 sum $k_2 \oplus k_3 \oplus k_4 \oplus c_3$. This equals zero if no errors have occurred during transmission. The output from the three EX-OR gates thus represents the syndrome which is fed into some combinational logic circuitry to determine the position of any error. The error is corrected by means of a k-stage shift register at the output of the combinational logic circuit. This will contain all zeros apart from any erroneous position which will contain a 1. The output from the shift register is added (modulo-2) serially to the received information bits and any bit position within the received data which gets added to a 1 will change from 0 to 1 (or 1 to 0), thus correcting the error. Unfortunately, circuits such as these, although simple in the case of the Hamming (7,4) code, become excessively complicated for the longer block lengths used in data communications networks. Hamming codes do, however, find uses in situations which do not require large block lengths, such as remote control of robotic systems.

6.3.2 Cyclic codes

As mentioned above, simple linear codes such as the Hamming codes have a limitation in that if large block lengths are used, for example in data communications, then the encoding and decoding circuitry becomes very complex. Paradoxically, the circuitry can be made simpler if the mathematical structure of the code is made more complex. A cyclic code is one in which all the code words are related by the fact that if a code word is rotated, it becomes another code word. The following code is obtained from the single check equation $c_1 = k_1 \oplus k_2$ and, although trivial, is cyclic:

$$k_1 \ k_2 \ c_1$$
$$0 \ 0 \ 0$$
$$0 \ 1 \ 1$$
$$1 \ 0 \ 1$$
$$1 \ 1 \ 0$$

All four of these code words can be rotated in either direction and will result in another code word. Consequently, to define this code, it is only necessary to have one nonzero code word, since all the other code words can be obtained from it (the all-zeros code word is obtained by adding any code word to itself). Cyclic codes are usually defined by a single code word expressed as a polynomial, known as a **generator polynomial**. For example, the cyclic code used in the ITU-T X.25 protocol has a generator polynomial $x^{16} + x^{12} + x^5 + 1$, where $x = 2$ since the code is binary. This is expressed as a polynomial rather than the binary number 10001000000100001 because the latter is rather unmanageable. The highest power of a generator polynomial is called its **degree** and is always equal to the number of check bits in the code. Since cyclic codes are invariably linear block codes, they can also be described by a generator matrix, which can be readily obtained from the generator polynomial as Example 6.4 shows.

EXAMPLE 6.4

A cyclic block code with block length $n = 7$ has a generator polynomial $x^3 + x + 1$. Determine the generator matrix, and the full set of code words.

The code words have 3 check bits since the degree of the polynomial is 3 and the number of information bits, $k = 7 - 3 = 4$. The generator polynomial, $G(x) = x^3 + x + 1 = 1011$ in binary. To make this into a 7-bit code word, 3 insignificant zeros are added to give 0001011. This code word is made the bottom row of the generator matrix and the other rows are obtained by rotating this code word to the left. The generator matrix is of size $k \times n = 4 \times 7$ as follows:

$$G = \begin{bmatrix} 1 & 0 & 1 & 1 & 0 & 0 & 0 \\ 0 & 1 & 0 & 1 & 1 & 0 & 0 \\ 0 & 0 & 1 & 0 & 1 & 1 & 0 \\ 0 & 0 & 0 & 1 & 0 & 1 & 1 \end{bmatrix} \quad \begin{matrix} a \\ b \\ c \\ d \end{matrix}$$

However, as we saw with Hamming codes, it is convenient if a generator matrix has an echelon to the left, when it is in standard echelon form. Rows c and d fit this pattern, but the top two rows need to be changed by means of modulo-2 addition as follows:

$$G = \begin{bmatrix} 1 & 0 & 0 & 0 & 1 & 0 & 1 \\ 0 & 1 & 0 & 0 & 1 & 1 & 1 \\ 0 & 0 & 1 & 0 & 1 & 1 & 0 \\ 0 & 0 & 0 & 1 & 0 & 1 & 1 \end{bmatrix} \quad \begin{matrix} a \oplus c \oplus d \\ b \oplus d \\ c \\ d \end{matrix}$$

Further code words can be obtained by further left rotations as follows:

0001011	0110001
0010110	1100010
0101100	1000101
1011000	

To obtain further code words it is necessary to add two together; choosing the last two above gives 1100010 + 1000101 = 0100111. Shifting of this code word provides the following:

0100111	1110100
1001110	1101001
0011101	1010011
0111010	

Two further code words are required to make a total of 16; one is the all-zeros code word 0000000 and the other is obtained by adding two code words as follows:

$$
\begin{array}{r}
0100111 \\
\oplus\ \underline{1011000} \\
1111111
\end{array}
$$

The fact that the code words of a cyclic code are obtained by shifting and adding the generator polynomial leads to the important characteristic that all code words are a multiple of the generator polynomial. To encode incoming information bits, a cyclic encoder must therefore generate check bits which, when added to the information bits, will produce a code word which is a multiple of $G(x)$, the generator polynomial. This is achieved as follows: first, the information bits, normally a lengthy bit pattern, are represented as a polynomial $K(x)$, where x is, in practice, 2. Second, let the information bits $K(x)$, followed by c zeros (that is, a code word with all the check bits set to zero) be represented by the polynomial $F(x)$ which is in fact $K(x)$ shifted by c places, that is, $x^c K(x)$. If $F(x)$ is now divided by the generator polynomial $G(x)$ then:

$$\frac{F(x)}{G(x)} = Q(x) + \frac{R(x)}{G(x)}$$

where $Q(x)$ is the quotient and $R(x)$ the remainder, that is:

$$F(x) = Q(x)\,G(x) + R(x)$$

If the remainder $R(x)$ is now added (modulo-2) to $F(x)$, we obtain:

$$F(x) + R(x) = Q(x)\,G(x)$$

since addition and subtraction will give the same result in modulo-2. It is this bit sequence, $F(x) + R(x)$, which is transmitted, since it is always a multiple of the generator polynomial $G(x)$ and is therefore always a legal code word.

Thus, encoding for a cyclic code consists of adding c zeros to the end of the information bits and dividing by the generator polynomial to find the remainder. (Note that modulo-2 division is used.) The remainder is added to the information bits in place of the c zeros and the resulting code word transmitted. At the receiver, a decoder tests to see whether the received bit sequence is error free by dividing again by the generator polynomial. An error-free transmission results in a zero remainder. This process is illustrated by Example 6.5.

EXAMPLE 6.5

A (7,4) cyclic code has a generator polynomial $x^3 + x + 1$. Information bits consisting of 1100 (most significant bit on the left) are to be coded and transmitted. Determine:

(a)   the transmitted code word,

(b)   the remainder obtained at the receiver if the transmission is error free,

(c)   the remainder obtained at the receiver if an error occurs in bit 4.

(a)   The generator polynomial $x^3 + x + 1$ expressed as a binary number is 1011. First, three zeros are added to the information bits to produce $F(x) =$ 1100000. Dividing $F(x)$ by the generator polynomial $G(x)$ using modulo-2 division gives:

```
            1110     Quotient
     1011 1100000
            1011
            1110
            1011
            1010
            1011
             010     Remainder
```

Thus the remainder = 010 and this is added to the information bits in place of the three zeros to give the transmitted code word 1100010.

(b)   If the transmission is error free, we divide the received data by $G(x)$ and obtain:

```
            1110
     1011 1100010
            1011
            1110
            1011
            1011
            1011
             000     Zero remainder
```

(c) If an error occurs in bit 4, the received bit pattern is 1101010, resulting in:

```
          1111
1011 ⟌1101010
      1011
      1100
      1011
       1111
       1011
       1000
       1011
        011      Remainder
```

The fact that this remainder in (c) is nonzero indicates that an error has occurred. Note that the division used is modulo-2 division and no 'carries' or 'borrows' can be used. In this example, the numbers that are used within the division process are restricted to four bits.

What if we now use a check matrix to determine the syndrome? A cyclic code is a linear block code and so it should be possible to do this. The check matrix for this code can be obtained from the generator matrix of Example 6.4 in exactly the same way as the check matrix was obtained in Example 6.2, as follows:

$$G = \begin{bmatrix} 1\,0\,0\,0\,1\,0\,1 \\ 0\,1\,0\,0\,1\,1\,1 \\ 0\,0\,1\,0\,1\,1\,0 \\ 0\,0\,0\,1\,0\,1\,1 \end{bmatrix} \quad \text{and} \quad H = \begin{bmatrix} 1\,1\,1\,0\,1\,0\,0 \\ 0\,1\,1\,1\,0\,1\,0 \\ 1\,1\,0\,1\,0\,0\,1 \end{bmatrix}$$

and the syndrome is obtained by multiplying the check matrix by the received data:

$$\begin{bmatrix} 1\,1\,1\,0\,1\,0\,0 \\ 0\,1\,1\,1\,0\,1\,0 \\ 1\,1\,0\,1\,0\,0\,1 \end{bmatrix} \times \begin{bmatrix} 1 \\ 1 \\ 0 \\ 1 \\ 0 \\ 1 \\ 0 \end{bmatrix} = \begin{bmatrix} 0 \\ 1 \\ 1 \end{bmatrix}$$

which gives the same result as the remainder obtained from the division. Although the two mathematical processes appear to be different, they produce the same result.

Modulo-2 division circuits

Modulo-2 division is achieved in an electronic circuit by repeated shifting and subtraction. This can be very easily implemented using shift registers and, bearing in mind that modulo-2 subtraction is identical to addition, EX-OR gates. A circuit that will divide by $1 + x + x^3$ is given in Figure 6.4:

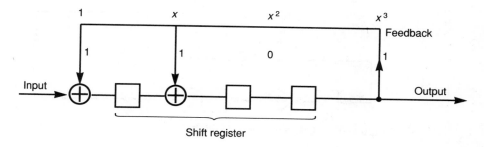

Figure 6.4 Modulo-2 division circuit.

The best way to understand the working of this circuit is to analyse its step-by-step operation. Firstly we choose an input bit pattern which will give a known remainder. In Table 6.1, the input 1100010 is used. We already know from Example 6.5 that this is divisible by 1011 ($x^3 + x + 1$) and should therefore produce a remainder of zero. Note that after six shifts, the shift register contains the required remainder of 000. The quotient will appear at the output of the circuit. However, in coding circuits, it is only the remainder that is of significance.

Table 6.1 Modulo-2 division.

Step J	Input on Jth shift	Feedback on Jth shift	Shift register after Jth shift	Output after Jth shift
0	–	–	0 0 0	0
1	1	0	1 0 0	0
2	1	0	1 1 0	0
3	0	0	0 1 1	1
4	0	1	1 1 1	1
5	0	1	1 0 1	1
6	1	1	0 0 0	0
7	0	0	0 0 0	

Cyclic encoding and decoding circuits

Encoding for a cyclic code consists of dividing the polynomial $F(x)$ (incoming information bits with zeros in the check bit positions) by the generator polynomial $G(x)$ to find a remainder $R(x)$ which is then added to $F(x)$ to give the code word $F(x) + R(x)$. To obtain the zero check bits, the information bits $K(x)$ are shifted c places prior to the division process. An encoding circuit, known as a Meggitt encoder, that achieves this for $G(x) = 1 + x + x^3$ is shown in Figure 6.5.

Bringing the input information bits $K(x)$ into the circuit at the point shown rather than to the left of the circuit as in the division circuit of Figure 6.4 is

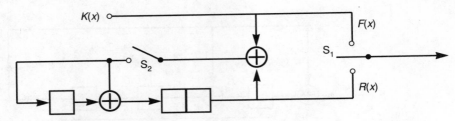

Figure 6.5 Cyclic encoding circuit.

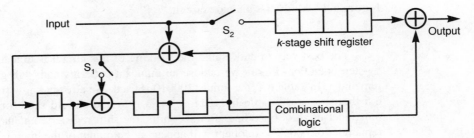

Figure 6.6 Cyclic decoding circuit.

equivalent to shifting c places ($c = 3$ in this case). Initially S_1 is in the up position and S_2 is closed, while $F(x)$ is simultaneously transmitted and processed in the division circuit. After k shifts, the shift register will contain the remainder $R(x)$. S_1 then moves to the down position and S_2 opens so that the remainder is then transmitted.

A similar arrangement is shown in the decoder circuit of Figure 6.6. Initially the information bits are received into a k-stage shift register with S_1 and S_2 closed. At the same time, the check bits are recalculated by the division circuit and then fed into a combinational logic circuit. The check bits are received with S_1 and S_2 open and are fed straight into the combinational logic circuit where they are compared with the recalculated check bits to determine a syndrome. S_1 and S_2 are then closed again and the syndrome is used by the combinational circuit to correct the data at the output of the decoding circuit.

6.4 Feedback error control

Even very powerful error-correcting codes may not be able to correct all errors that arise in a communication channel. Consequently, many data communications links provide a further error-control mechanism, in which errors in data are detected and the data is retransmitted. This procedure is known as **feedback error control** and it involves using a channel code to detect errors at the receive end of a link and then returning a message to the transmitter requesting the retransmission of a block (or blocks) of data. Alternatively, errors in received data can be detected and corrected up to a certain number of errors and a retransmission requested only if more than this number of errors occurs. The process of retransmitting the data is also known as **Automatic Repeat Request** (ARQ). There are three types of ARQ in use **stop-and-wait**, **go-back-*n*** and **selective-repeat**.

6.4.1 Stop-and-wait ARQ

This technique, which is also known as **idle RQ**, is the simplest method of operating ARQ and ensures that each transmitted block of data or **frame** is correctly received before sending the next. At the receiver, the data is checked for errors and if it is error free an acknowledgement (ACK) is sent back to the transmitter. If errors are detected at the receiver a negative acknowledgement (NAK) is returned. Since errors could equally occur in the ACK or NAK signals, they should also be checked for errors. Thus, only if each frame is received error free and an ACK is returned error free can the next frame be transmitted. If, however, errors are detected either in the transmitted frame or in the returned acknowledgement, then the frame is retransmitted. A further (and hopefully remote) possibility is that a frame or acknowledgement becomes lost for some reason. To take account of this eventuality, the transmitter should retransmit if it does not receive an acknowledgement within a certain time period known as the **timeout interval**. Finally, a frame may be received correctly but the resulting ACK may be lost, resulting in the same frame being transmitted a second time. This problem can be overcome by numbering frames and discarding any correctly received duplicates at the receiver. Although stop-and-wait offers a simple implementation of an ARQ system it can be inefficient in its utilization of the transmission link since time is spent in acknowledging each frame. Efficiency of utilization of a link is an important consideration in the design of a communications link which will be explored in Section 7.1.1.

6.4.2 Go-back-*n*-ARQ

If the error rate on a link is relatively low, then the link efficiency can be increased by transmitting a number of frames continuously without waiting for an immediate acknowledgement. This strategy is used in go-back-*n* ARQ which is used in a number of standard protocols including HDLC (High-Level Data Link Control). HDLC will be covered in some detail in Section 7.4. The go-back number *n* determines how many frames can be sent without an acknowledgement having been received.

Frame $i + n$ cannot be transmitted until frame i has been acknowledged. Figure 6.7 shows a typical transfer of frames for a go-back-4 ARQ system over a full-duplex link.

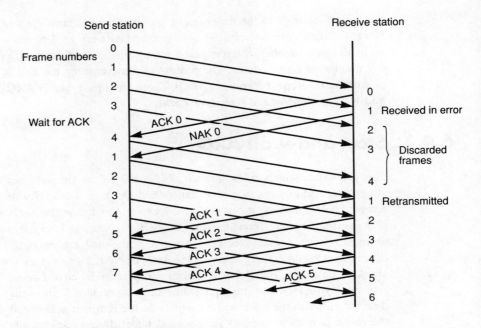

Figure 6.7 Frame transfer go-back-*n*.

Note that the send station transmits four frames (numbered 0, 1, 2, 3) without receiving an acknowledgement. There then follows a short wait before the receipt of the first ACK which contains the number of the last correctly received frame (0 in this case). Although this acknowledgement is returned immediately after the correct receipt of frame 0, transmission delays mean that it does not reach the send station until after frame 3 has been transmitted (the send station having carried on transmitting as a result of the go-back-4 strategy). If we now assume that an error occurs in frame 1; the receive station will reply with NAK 0, indicating that an error has been detected and that the last correctly received frame was frame 0. The send station will now *go back* and retransmit the frame after the last correctly received frame, which in this case is frame 1. Meanwhile the receive station will have received frames 2, 3 and 4 which, since they have not been acknowledged, need to be discarded. In the absence of further errors, the transmit station continues transmitting frames 2, 3, 4, 5, ... for as long as ACK signals continue to be returned. A go-back-*n* strategy allows for the efficient transfer of frames without the need for any substantial buffer storage by receive equipment as long as error rates are low. However, if errors occur during transmission then frames that have already been received correctly need to be retransmitted along with erroneous frames. If buffer storage is available at the receive equipment then it seems reasonable that some form of selective retransmission strategy would be more efficient than go-back-*n* in the presence of errors.

6.4.3 Selective-repeat ARQ

In selective-repeat ARQ only those frames that generate a NAK are retransmitted. Although this appears more efficient than go-back-*n*, it requires sufficient storage at the transmitter to save all frames that have been transmitted but not acknowledged in case a frame proves to be erroneous. In this system a go-back number is still used to determine how many frames can be transmitted without receiving an acknowledgement. Selective-repeat ARQ has not been a particularly popular ARQ strategy because of the increased memory requirements mentioned but, as memory capabilities become more easily available, it offers potential efficiency gains in the presence of high error rates.

Exercises

6.1 Explain whether or not ACKs and NAKs need to be numbered in a stop-and-wait ARQ system.

6.2 By adding together rows (modulo-2 addition), rearrange the following generator matrix so that it is in standard echelon form:

$$G = \begin{bmatrix} 1\,0\,0\,1\,1 \\ 1\,1\,0\,0\,1 \\ 1\,1\,1\,0\,0 \end{bmatrix}$$

6.3 An (8,4) linear block code has the following check equations:

$$c_1 = k_2 \oplus k_3 \oplus k_4$$
$$c_2 = k_1 \oplus k_2 \oplus k_3$$
$$c_3 = k_1 \oplus k_2 \oplus k_4$$
$$c_4 = k_1 \oplus k_3 \oplus k_4$$

(a) Determine the generator and check matrices.

(b) What is the minimum distance of this code?

(c) Sketch an encoding circuit for this code.

6.4 By adding a fourth check bit to the Hamming (7,4) code of Section 6.3.1, devise an (8,4) code. For this code, determine:

(a) the check matrix,

(b) the generator matrix,

(c) the minimum distance of the code,

(d) the numbers of errors that can be detected and that can be corrected.

6.5   Divide the polynomial $x^7 + 1$ by $x^3 + x + 1$ (modulo-2 division) and hence determine the factors of $x^7 + 1$.

6.6   Devise a circuit that will divide by $x^3 + x + 1$ (modulo-2 division). If data consisting of the polynomial $x^6 + x^5 + x + 1$ is fed into the circuit (most significant bit first), determine the contents of the circuit shift register after the seventh shift.

6.7   A (7,3) cyclic code has a generator polynomial $x^4 + x^2 + x + 1$.

   (a)   Determine the generator and check matrices of this code.

   (b)   Sketch circuits of a Meggitt encoder and decoder for this code.

   (c)   Obtain the complete code set.

6.8   A cyclic code with block length 7 has the following generator polynomial:

$$G(x) = x^3 + x^2 + 1$$

Determine:

   (a)   the number of information and check bits in a code word,

   (b)   the number of code words in the code,

   (c)   the generator and check matrices of the code,

   (d)   the minimum distance of the code.

6.9   Data consisting of the sequence 10101010101 is to be encoded and transmitted using a cyclic (15,11) code with generator polynomial $x^4 + x^2 + x + 1$. Using modulo-2 division, determine the 4 check bits that need to be added before transmission. The encoded data is transmitted and an error occurs in bit 8. Determine the syndrome (remainder).

Link control and management

7

Chapters 3–6 were mainly concerned with the production of data signals suitable for reliable transmission over a data communications link. In particular, we looked at the different types of coding used: source codes in Chapter 3; line codes in Chapter 4; and channel codes in Chapter 6. Of these, channel codes and the more general topic of error control are covered by level two, the data link layer, of the OSI model. As mentioned in Chapter 1, this layer is also concerned with the control and management of a data communications link and has two further functions, namely link management and flow control. Link management involves the setting-up and disconnecting of a link. Flow control ensures that data in the form of frames is transmitted in an orderly fashion so that, for example, a send station does not transmit frames faster than a receive station can accept them. However, before moving on to these topics it is necessary to look at the measurement of the flow of data over a link.

7.1 Link throughput

The transmission rate of a data link, once established, represents an upper limit for the transfer of information over the link. In practice, a variety of factors cause the useful information transfer rate to be reduced to a value below that of the transmission rate. As mentioned in Section 2.1.3, most data communication systems divide data into fixed-length blocks or frames. We can define throughput as the number of information bits in a frame (or frames) divided by the total time taken to transmit and acknowledge the frame (or frames). The major factors that cause throughput to be less than the transmission rate are listed below as follows:

(1) *Frame overheads* Not all of the contents of a frame are information bits. Typically, in addition to information bits, a frame also contains a *header* and a *trailer*. The header contains control information such as an address and sequence numbers. The trailer contains error-checking bits which are often called a **frame check sequence**. A frame might typically contain 1072 bits of which only 1024 are information bits, thus leading to a 5% reduction in potential throughput even before the frame is transmitted.

(2) *Propagation delay* This is the time that it takes for a frame to propagate from one end of a link to the other, that is, the difference in time between the first bit of a frame leaving the send station and arriving at the receive station. Propagation delay must not be confused with the frame transmission time which is the difference in time between the first bit and the last bit of a frame leaving the send station. Propagation delay often has only a small effect on throughput but in some situations, such as long-distance radio links and especially satellite links, it can be a major factor in reducing the throughput if acknowledgements are used.

(3) *Acknowledgements* Normally, some form of ARQ is used and time may be spent waiting for acknowledgements to reach the send station, particularly if there is a half-duplex link. Since the acknowledgements will normally be much shorter than the information frames, the transmission time of the acknowledgements can often be ignored. However, the propagation delay of an acknowledgement will be the same as that of an information frame providing they take the same transmission path.

(4) *Retransmissions* Frames may need to be retransmitted as a result of errors. The retransmission is accompanied by acknowledgements if ARQ is being used. If the error rate is high then this is the most serious cause of reduction in throughput.

(5) *Processing time* Time is spent at the send and receive stations in processing the data. This includes detecting (and possibly correcting) errors and also the implementation of flow control. If modems are used there will be further processing delays associated with the modulation and demodulation processes.

7.1.1 Link utilization

The best way to illustrate the concept of throughput is by way of an example. However, before we look at an example, it is useful to define a further term, known as **link utilization** or **efficiency**. Link utilization is the ratio of the time taken to transmit a frame or frames of data to the total time it takes to transmit and acknowledge the frame or frames:

$$\text{Utilization, } U = \frac{\text{Time taken to transmit frame}}{\text{Total transmission time}}$$

Link utilization also depends on the type of ARQ used. The utilization of a link with stop-and-wait ARQ can be determined as follows. If the time taken to transmit a frame or block of data is t_f, the propagation delay for both frame and acknowledgement is t_d, the time taken to transmit an acknowledgement is t_a and the total processing time is t_p, then:

$$U = \frac{t_f}{t_f + t_a + t_p + 2t_d}$$

In many situations the acknowledgement transmission time and processing times can be ignored, giving:

$$U = \frac{t_f}{t_f + 2t_d} = \frac{1}{1 + 2a} \qquad \text{where } a = t_d/t_f$$

The concept of utilization is often used in the analysis of local area networks and is also explored in some detail in Section 9.2.

EXAMPLE 7.1

A half-duplex point-to-point satellite transmission link connecting two computers uses a stop-and-wait ARQ strategy and has the following characteristics:

> Data transmission rate = 4.8 kbps
> Frame size, $n = 2040$ bits
> Information bits per frame, $k = 1920$
> Propagation delay, $t_d = 250$ ms
> Acknowledgement size = 48 bits
> Round-trip processing delay, $t_p = 25$ ms

Determine the throughput and link utilization.

> Frame transmission time, $t_f = \dfrac{2040}{4800} = 0.425$ s
>
> Acknowledgement transmission time, $t_a = \dfrac{48}{4800} = 10$ ms
>
> Total time to transmit frame and receive an acknowledgement
> $= t_f + t_a + t_p + 2t_d = 0.425 + 0.01 + 0.025 + 0.5 = 0.96$ s
>
> Throughput $= k = \dfrac{1920}{0.96} = 2.0$ kbps

Note that the resulting throughput is considerably less than the transmission rate of 4.8 kbps. If the values of t_a and t_p had been neglected the value of the throughput would have been calculated as:

> $\dfrac{1920}{0.925} = 2.076$ kbps

The link utilization can now be calculated, neglecting t_a and t_p, as follows:

> $a = t_d/t_f = 0.25/0.425 = 0.588$
>
> $\dfrac{1}{1 + 2a} = \dfrac{1}{1 + 1.176} = 46\%$

7.1.2 Effect of errors on throughput

As mentioned in Section 6.1, the effect of transmission impairments on a data communication link is to introduce errors. As mentioned in Section 2.1.7 the number of errors present in a link is expressed as a bit error rate (BER). If a link has a BER of 0.00001 (10^{-5}), this means that there is a probability of 0.00001 that any bit is in error. Alternatively, we can say that, on average, one in every 100 000 bits will be in error. This may seem a very low error rate but if bits are transmitted as a block in a frame then the probability of the frame being in error will be much greater. The frame error rate, P, can be obtained from the bit error rate, E, as follows: The probability of a bit being error free is $1 - E$ and the probability of a block of length n being error free is $(1 - E)^n$. The frame error probability is therefore:

$$P = 1 - (1 - E)^n$$

EXAMPLE 7.2

A frame of data of length 2048 bits is transmitted over a link with a BER of 10^{-4}. Determine the probability that a frame will be received erroneously.

Bit error rate, $E = 0.0001$. Probability of a bit being error free = $1 - 0.0001 = 0.9999$. If the frame length, n, is 2048 bits then the probability of the frame being error free is $(0.9999)^{2048} = 0.815$.

The probability of a frame being in error is given by:

$$P = 1 - 0.815 = 0.185$$

so that even though only one in every 10 000 bits is, on average, in error there is an almost 20% chance of a frame being received in error and almost 1 in every 5 frames will need to be either corrected at the receiver or retransmitted.

7.1.3 Effect of ARQ on throughput

The situation illustrated in Example 7.2 is compounded if ARQ is used since retransmitted frames are equally likely to contain errors (errors in acknowledgements are much less likely since they normally are of very short length). Thus a frame may need to be retransmitted a number of times and the probability of retransmission will be equal to the probability of a frame containing errors since only erroneous frames are retransmitted. If a frame is retransmitted m times then the probability of this occurring is the probability of transmitting m consecutive erroneous frames followed by a single correctly received frame. This is given by:

$$P^m(1 - P)$$

A problem which arises is that the number of times, m, that a frame is retransmitted will vary according to some form of probability distribution. Since the determination of the value of m is not particularly simple, the value is just presented here without any analysis. For a full analysis see Bertsekas and Gallager (1987). If stop-and-wait ARQ is used then the average number of times that a frame is transmitted is given by:

$$m = 1 + \frac{P}{1-P} = \frac{1}{1-P}$$

If go-back-n ARQ is used then an error detected in a frame causes that frame, along with all other unacknowledged frames, to be retransmitted. Since this situation is more complicated than with a stop-and-wait strategy, we shall make the following assumptions to simplify the calculation:

(1) Frames are retransmitted only when a frame is rejected at the receiver for being erroneous. In practice, there may be other reasons for frames being retransmitted.

(2) The rejection of frame i by the receiver is followed by the transmitter sending frames $i + 1$, $i + 2$, ..., $i + n - 1$ and then retransmitting the original frame i. This may not always be the case since there may be fewer than $n - 1$ frames waiting to be transmitted after frame i.

The resulting analysis, which is also carried out in Bertsekas and Gallager (1987), gives the number of times that a frame is likely to be transmitted as:

$$m = 1 + \frac{nP}{1-P}$$

This result will be used in Example 7.5. Selective-repeat will produce the same result as stop-and-wait in respect of the number of times that a frame is transmitted. However, selective-repeat normally outperforms stop-and-wait significantly in terms of the throughput. To appreciate the effect of ARQ on throughput consider Example 7.3.

EXAMPLE 7.3

A long-distance radio link uses a stop-and-wait ARQ strategy with half-duplex transmission and has the following characteristics:

> Data transmission rate = 5.4 kbps
> Frame size, $n = 896$
> Information bits per frame, $k = 784$
> Propagation delay, $t_d = 10$ ms

If processing delays and acknowledgement transmission time can be neglected, determine the throughput: (a) in the absence of errors and (b) in the presence of a bit error rate of 10^{-3}.

(a)   Frame transmission time:

$$t_f = \frac{896}{5400} = 0.166\,s$$

Throughput in the absence of errors:

$$\frac{k}{t_f + 2t_d} = \frac{784}{0.166 + 0.02} = 4217\,bps$$

(b)   Frame error rate:

$$P = 1 - (1 - E)^n = 1 - (1 - 0.001)^{896} = 0.59$$

Average number of times a frame is transmitted:

$$1 + \frac{P}{1 - P} = 1 + \frac{0.59}{0.41} = 2.44$$

The transmission and delay times will be increased by this amount, giving a throughput:

$$\frac{k}{2.44(t_f + 2t_d)} = \frac{784}{2.44(0.166 + 0.02)} = 1727\,bps$$

The effect of the errors in part (b) of Example 7.3 is to reduce the throughput to less than one-third of the data transmission rate. This reduction in throughput is fairly excessive and is a result of the high BER of 10^{-3}. A good quality data communications link is more likely to have a BER of 10^{-8}.

7.1.4 Optimum block length

What is the effect of altering the length of a frame in Example 7.3? If the frame length is increased in an error-free situation then the throughput increases since the value of k is greater and the value of t_d remains the same. However, in the presence of errors the situation is more complicated. An increased block size still produces a larger number of information bits transmitted in each block but a point will be reached at which throughput falls as a result of having to retransmit a large block of data each time an error is detected. This leads us to consider an optimum length of block. Figure 7.1 shows a plot of throughput against block length for the system described in Example 7.3 part (b), with a BER of 10^{-3}.

Note that the throughput is about 1720 bps for a block length of 900, as expected from the result of the problem. Note also that there is a maximum value of throughput of about 2130 bps which occurs at a block length of 490 bits. Figure 7.2 shows what happens if the BER deteriorates even further to 10^{-2}.

The maximum throughput has been reduced to 230 bps and now occurs at a block length of 180 bits. There are now so many blocks requiring retransmission that the only way to maintain any throughput is to restrict the blocks to this short length.

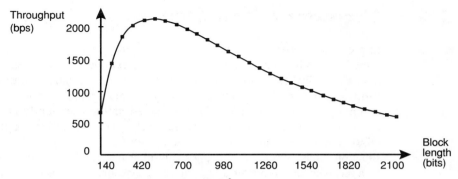

Figure 7.1   Throughput for BER of 10^{-3}.

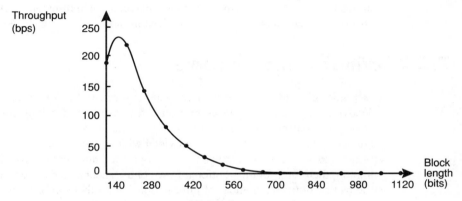

Figure 7.2   Throughput for BER of 10^{-2}.

The choice of block length is particularly important if data is being transmitted long distances in the presence of a high BER. A typical application in which it is crucial to choose an appropriate block length is radio-frequency modems operating over distances of thousands of kilometres.

7.2 Flow control

Whichever form of ARQ is used over a data communications link, a certain amount of buffer storage is required at both ends of a link to allow for the processing of data. Even in a character-orientated link there needs to be buffer storage for at least one character. Flow control involves the control of the data transmission so that transmitters and receivers have sufficient buffer space to process each frame or character as and when it arrives. Computers in a time-share system, for example, may not always be able to process the characters transmitted to them at the maximum bit rate. Two flow control schemes are in common use depending on the type of link and these are described in Sections 7.2.1 and 7.2.2.

7.2.1 Character-oriented link

In a character-oriented asynchronous link, such as a DTE communicating with a remote computer, characters are sometimes **echoed** back as a simple check to determine whether or not they have been received correctly. (This is a primitive form of error checking.) The problem with this simple technique is that if the computer becomes temporarily overloaded and consequently ceases to echo characters then there is no automatic mechanism for preventing further transmission and the system relies on human intervention to stop transmission. This situation can be improved by using a control character, X-OFF, which is transmitted by a receiver (the computer in this case) if it becomes overloaded. On receipt of the X-OFF character, the transmitter (DTE) will cease transmitting, possibly buffering any further characters. When sufficient buffer space becomes available at the computer to allow the receipt of further characters, it returns an X-ON character. This instructs the transmitter to resume transmission.

7.2.2 Window mechanisms

If a frame-oriented system uses a continuous form of ARQ such as go-back-*n* or selective-repeat then, as long as information is available for transmission, a send station can continue sending information frames before receiving an acknowledgement. The send station will be provided with a predetermined amount of buffer storage and it is important that this storage does not become overloaded. It is common, therefore, for an additional flow control mechanism to be used with these systems that limits the number of information frames that can be transmitted before receiving an acknowledgement. The send station keeps a copy of those frames transmitted but not acknowledged so that it can retransmit if necessary. A maximum limit is set on the number of copies that are being held at the send station which is known as the **send window**. If the send station reaches its maximum window size it stops transmitting and, in the absence of any acknowledgements, it does not transmit any more frames. When the send station finally receives an acknowledgement it can start transmitting again. The window size is chosen so that it does not impede the flow of frames. As well as the amount of send buffer storage available, the frame size and transmission rate are also taken into account in determining the window size. The operation of a send window is illustrated in Figure 7.3.

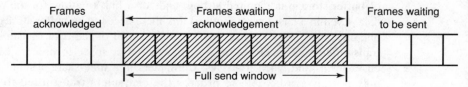

Figure 7.3   Operation of send window: (a) window full.

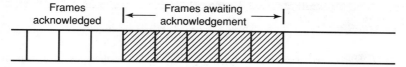

Figure 7.3 *continued*   Operation of send window: (b) continuous flow possible.

A list of information frames transmitted but, as yet, unacknowledged is kept at the send station. Each time an information frame is transmitted the list is increased by one and each time an acknowledgement is received it is reduced by one.

7.2.3 **Effect of windows on throughput**

As we saw in Example 7.1, the throughput of a frame-oriented link using stop-and-wait ARQ depends upon the number of information bits, k, the frame transmission time, t_f, and the propagation delay, t_d, according to the expression:

$$\text{Throughput} = \frac{k}{1 + 2a} \qquad \text{where } a = \frac{t_d}{t_f}$$

This assumes that, as in Example 7.3, both the processing delays and acknowledgement transmission times can be neglected. Now, if a send window of size N is used, this will clearly have some effect on the throughput. Consider a send station commencing transmission of a stream of data at time t_o. The leading edge of the first data bit will reach the receive station at time $t_o + t_d$ and the trailing edge of the last data bit will reach the receive station at time $t_o + t_d + t_f$. Neglecting the processing time means that we can assume that the receive station responds immediately by transmitting an acknowledgement. The acknowledgement (which we have already assumed to be of negligible length) will reach the send station at time $t_o + 2t_d + t_f$. There are now two possible scenarios depending on the window size N:

(1)   The acknowledgement transmitted by the receive station reaches the send station before the send window is full. This can be represented by the expression:

$$Nt_f > 2t_d + t_f$$

In this case the send station is able to transmit continuously without having to stop as a result of the send window becoming full. In the absence of errors, throughput is optimal.

(2)   The send window becomes full at time $t_o + Nt_f$. The send station then stops transmitting until an acknowledgement is returned that reaches the send station at time $t_o + 2t_d + t_f$, at which point the send station may resume transmission. This situation can be represented by the expression:

$$Nt_f < 2t_d + t_f$$

Clearly, the throughput is restricted under these circumstances.

EXAMPLE 7.4

A frame-oriented data communications system operates at a transmission rate of 96 kbps with a frame length of 1024 bits over a long-distance link which produces a propagation delay of 20 ms. A flow control system is required using a window mechanism. Determine the minimum window size which allows for optimum throughput.

Optimum throughput cannot be achieved unless the expression of Section 7.2.3(1) is satisfied as follows:

$$Nt_f > 2t_d + t_f$$

$$N > \frac{2t_d}{t_f} + 1$$

The delay time, $t_d = 20$ ms.
The frame transmission time:

$$t_f = \frac{\text{frame length}}{\text{bit rate}} = \frac{1024}{96\,000} = 10.667 \text{ ms}$$

Substituting these times gives a value for N:

$$N > 2 \times \frac{20}{10.67} + 1 = 4.75$$

Thus the minimum window size for efficient operation of the link is 4.75 and a window size of 7 would be adequate, requiring 3 bits to specify a frame within the window at any time.

It is useful at this stage to determine the utilization of a link that uses windows in conjunction with some form of continuous ARQ (go-back-n or selective-repeat). If N frames are transmitted continuously then the time spent in transmitting data will be Nt_f. The value of the utilization will depend on which of the two scenarios mentioned above applies, as follows:

(1) If the send station is able to transmit continuously, the utilization is given by:

$$U = \frac{\text{time spent transmitting frames}}{\text{total time}} = \frac{Nt_f}{Nt_f} = 1$$

(2) If the send window becomes full after a time Nt_f, when transmission stops until an acknowledgement is received after time $t_f + 2t_d$, the expression becomes:

$$U = \frac{\text{time spent transmitting frames}}{\text{total time}} = \frac{Nt_f}{t_f + 2t_d} = \frac{N}{1 + 2a}$$

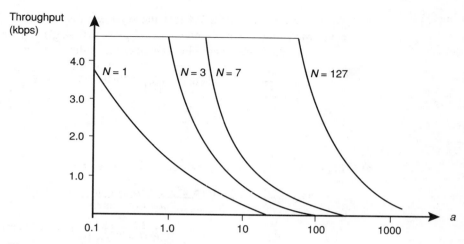

Figure 7.4   Effect of window size on throughput.

The effect of window size on utilization is shown in Figure 7.4. Throughput is plotted against $a = t_d/t_f$ for different values of window size, N.

A window size of $N = 1$ corresponds to a stop-and-wait ARQ strategy which, although it never provides an optimum value of throughput, is adequate for relatively short transmission links in which the propagation delay is short. A window size of 7 is adequate for most terrestrial links but in the case of links which have large propagation delays, such as satellite systems, a large window size is required to produce a satisfactory throughput (127 being a typical window size for such systems).

EXAMPLE 7.5

If the data communication system of Example 7.4 operates using a go-back-3 ARQ system (with a window size of 3) and 992 information bits in each frame, determine:

(a)   the utilization,

(b)   the throughput obtained on an error-free link,

(c)   the throughput for a BER of 10^{-4}.

Frame length, $n = 1024$
Information bits, $k = 992$
Bit rate = 96 kbps
Delay time, $t_d = 20$ ms
Frame transmission time, $t_f = 10.667$ ms
Window size, $N = 3$

(a) We know from Example 7.4 that the send window becomes full before an acknowledgement is received (since N is less than 4.75) and the utilization is not optimized. Under these circumstances, the utilization is given by:

$$U = \frac{N}{1 + 2a} \qquad \text{where } a = t_d/t_f = 20/10.667 = 1.875$$

Therefore:

$$U = \frac{3}{4.75} = 0.63$$

(b) In the absence of errors, the throughput is given by:

$$\text{Throughput} = \frac{\text{number of information bits transmitted}}{\text{total time taken}}$$

$$= \frac{Nk}{t_f + 2t_d} = \frac{3 \times 992}{(10.667 + 40) \times 10^{-3}} = 58.74 \text{ kbps}$$

(c) Bit error rate, $E = 10^{-4} = 0.0001$
Frame error probability is given by:

$$P = 1 - (1 - E)^n = 1 - (1 - 0.0001)^{1024} = 0.1$$

The effect of errors and the go-back-3 ARQ strategy will be for frames to be retransmitted. The average number of retransmissions, m, is given in Section 7.1.3. as:

$$m = 1 + \frac{NP}{1 - P} = 1 + \frac{(3 \times 0.1)}{0.9} = 1.33$$

The total time taken by the data transfer will be increased by this amount, giving a value for throughput as follows:

$$\text{Throughput} = \frac{Nk}{m(t_f + 2t_d)} = \frac{58.74}{1.33} = 44 \text{ kbps}$$

The throughput has been reduced to under half the transmission rate. This is partly due to the presence of errors, but more important is the fact that an inappropriate window size has been chosen. If a window size of, for example, 7 had been chosen then the throughput would have been optimized. Many systems allow for the window size to be configured in software and an appropriate choice of window size is an important consideration for network engineers.

The discussion above concerns data transmission in one direction only. If two-way data transmission is used then a window is required at each end of the link. In this case a technique known as **piggybacking** is often used, in which an acknowledgement signal is returned inside an information frame. In links in which propagation delay is high, piggybacking improves the link throughput considerably since separate acknowledgements can be dispensed with.

7.3 Link management

The flow control that has been discussed so far and the error control techniques of Chapter 6 are both concerned with the transfer of data across a link. Flow control ensures that the data is in the correct sequence and error control that the data is received correctly without errors. However, before either of these functions can take place the link needs to be set up and after the data transfer has taken place it needs to be disconnected. These two functions are known as **link management**. In the case of a physically short link these functions can be carried out by using separate control lines over which handshaking signals can be exchanged. An example of such a procedure is the ITU-T V.24 protocol mentioned in Chapter 5. In frame-oriented links, which are normally established over longer distances, it is normal to exchange separate **supervisory** frames over the same channel as the information frames. Clearly, these frames, as in the case of acknowledgements, need only be of a short length compared with the information frames. Figure 7.5 shows the signal flow diagram of a typical link set-up and disconnection procedure. As can be seen, two supervisory frames are used, namely a SETUP-frame and a DISC-frame. On the transmission of the SETUP-frame, frame numbers are set to zero and send and receive windows are initialized. Note that the supervisory frames need to be acknowledged since they may be corrupted by errors. Once the link is established, information frames and acknowledgements can be exchanged. Once the data transfer has ended a DISC-frame is used to terminate the logical connection between the two stations.

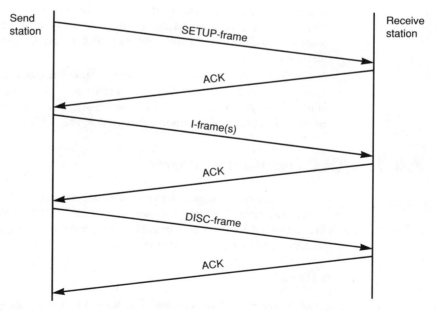

Figure 7.5 Link set-up and disconnection.

This process seems almost trivial at first sight but the situation becomes more complex if a failure occurs on a link or at a node. A problem arises when frames have been accepted for transmission over a link but have not reached a receive station before a failure occurs. Link management procedures need to be able to cope with such failures.

7.4 High-level data link control protocol (HDLC)

HDLC is a commonly used protocol developed by the ISO which is used to control data transfer over a link and includes not only the functions of flow control and link management already referred to in this chapter and in previous sections but also error control. It thus serves as a good practical illustration of the principles discussed in this chapter. The protocol allows for a variety of different types of link. To satisfy the requirements of different types of link the protocol distinguishes between three modes of operation (although only two of them are normally used) and two types of link configuration:

(1) *Unbalanced configuration* This is the situation in which a single **primary** station has control over the operation of one or more **secondary** stations. Frames transmitted by the primary are called **commands** and by a secondary **responses**. A typical example of this type of configuration is a **multidrop link** in which a single computer is connected to a number of DTEs which are under its control. This mode of working is called **normal response mode** (NRM). HDLC also specifies an alternative mode, called asynchronous response mode, for use within unbalanced configurations which, since it is rarely used, is not explained here.

(2) *Balanced configuration* This refers to a point-to-point link in which the stations at each end of the link have equal status. HDLC calls these **combined stations** and they can transmit both commands and responses. This mode of working is called **asynchronous balanced mode** (ABM).

7.4.1 HDLC frame structure

HDLC uses synchronous transmission with data being transmitted in frames. All frames have the common format shown in Figure 7.6. The address and control fields are known collectively as a **header** and the error-checking bits are called the Frame Check Sequence (FCS) or **trailer**.

Flag fields

These two fields mark the start and finish of the frame with the bit sequence 01111110. A single flag field may be used to mark the end of one frame and the

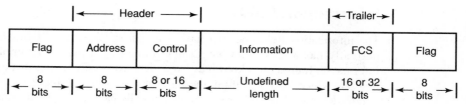

Figure 7.6   HDLC frame structure.

start of another if one frame follows immediately after another. Since receive stations are continuously looking for the flag sequence of six consecutive binary 1 bits to indicate the start or finish of a frame, it is essential that this sequence is prevented from occurring elsewhere in the frame. This is achieved by a process known as **bit stuffing** which takes place at the send station. When the frame is being assembled, if a sequence of five consecutive 1s appears within a frame then an extra 0 is inserted (stuffed) immediately after the sequence. At the receive station, when a sequence of five 1s is received, the next bit is examined. If this bit is a zero it is removed. If the sequence is followed by the two bits 10 then the combination is accepted as a flag. If the sequence is followed by two more 1s it is assumed that some form of error has occurred.

Address field

The contents of the address field depend on the mode of operation being used. In an unbalanced configuration it contains an 8-bit address which always identifies the secondary station, whether it is the primary or secondary that is transmitting. Alternatively, a group of secondary stations may have the same address, known as a **group address**, in which case a frame is transmitted from the primary station to all secondary stations in the group. The unique address containing all 1s is used to allow a primary to broadcast a frame to all the secondary stations connected to it. The protocol also allows for the length of the address field to be extended in the event of the number of secondaries being too large to be addressed by an 8-bit address. In a balanced configuration the address field always contains the address of the destination. Since a balanced configuration involves a point-to-point link, the destination address is not strictly required but is included for consistency. Note that the address is not used for any routing purposes since routing is a function of the network level of the ISO model and HDLC is primarily a link-level protocol.

Control field

The control field distinguishes between the three different types of frame used in HDLC, namely information, control and unnumbered frames. The first one or two bits of the field determine the type of frame. The field also contains control information which is used for flow control and link management.The function of this control information will be dealt with in more detail in Section 7.4.2.

Information field

The information field does not have a length specified by HDLC. In practice, it normally has a maximum length determined by a particular implementation. Information frames (also known as I-frames) are the only frames that carry information bits which are normally in the form of a fixed-length block of data of several kilobits length. All other types of frame normally have an empty information field.

Frame check sequence

The FCS field contains error-checking bits, normally 16 but with a provision for increasing this to 32 in the event of systems operating in an unreliable environment or with particularly long I-frames. The detail of the error-checking code used is covered in Section 7.4.4.

7.4.2 Frame types

The different types of frame are distinguished by the contents of the control field. The structure of all three types of control field is shown in Figure 7.7.

Information frames

An I-frame is distinguished by the first bit of the control field being a binary 0. Note also that the control field of an I-frame contains both a send sequence number, N(S), and a receive sequence number, N(R), which are used to facilitate flow control. N(S) is the number of frames sent and N(R) the number of frames successfully received by the sending station prior to the present frame being sent. Thus the first frame transmitted in a data transfer has send and receive sequence numbers 0,0. Since three bits are available for each of the sequence numbers N(S) and N(R), they can have values only between 0 and 7, that is, they use modulo-8 numbering. This imposes a limit on the size of the windows used for flow control. The use of send and receive sequence numbers is examined in more detail in Section 7.4.5. I-frames also contain a poll/final (P/F) bit (as do other frames). This acts as a poll bit when used by a primary station and a final bit by a secondary. A poll bit is set when a primary is transmitting to a secondary and requires a frame or frames to be returned in response and the final bit is set in the final frame of a response. Since there are no primaries or secondaries in asynchronous balanced mode, the P/F bit is used differently in this mode as we shall see later.

Frame type	1	2	3	4	5	6	7	8	Bits
Information	0		N(S)		P		N(R)		
Supervisory	1	0	F		P		N(R)		
Unnumbered	1	1	F		P		F		

N(S) = send sequence number, N(R) = receive sequence number, F = function bits,
P = poll/final bit used for polling in normal response mode

Figure 7.7 Control field structure.

Supervisory frames

Supervisory frames are distinguished by the first two bits of the control field being 10. These frames are used as acknowledgements for flow and error control. HDLC allows for both go-back-n and selective-repeat ARQ, although the latter has not been used extensively in the past because of the large memory buffering requirements. Note that the supervisory frames contain only a receive sequence number since they relate to the acknowledgement of I-frames and not to their transmission. They also contain two function bits which allow for four functions as shown in Table 7.1 which lists the supervisory commands/responses.

Table 7.1 Supervisory commands and responses.

Name	Function
Receive Ready (RR)	Positive acknowledgement (ACK), ready to receive I-frame
Receive Not Ready (RNR)	Positive acknowledgement, not ready to receive I-frame
Reject (REJ)	Negative acknowledgement (NAK), go-back-n
Selective Reject (SREJ)	Negative acknowledgement, selective repeat

Unnumbered frames

Unnumbered frames do not contain any sequence numbers (hence their name) and are used for various control functions. They have five function bits which allow for the fairly large number of commands and responses listed in Table 7.2.

Table 7.2 Unnumbered commands and responses.

Name	Function
Set Normal Response Mode (SNRM)	
Set Asynchronous Response Mode (SARM)	Used to initialize or change modes
Set Asynchronous Balanced Mode (SABM)	
Set Initialization Mode (SIM)	Used to initialize a link
Disconnect (DISC)	Causes the link connection to be terminated
Unnumbered Acknowledgement (UA)	Acknowledges the above mode-setting commands
Request Initialization Mode (RIM)	Requests SIM command when initialization required
Request Disconnect (RD)	Requests a disconnection of a link
Disconnected Mode (DM)	Indicates responding station disconnected
Unnumbered Poll (UP)	Used to request control information
Unnumbered Information (UI)	Used to exchange control information
Frame Reject (FRMR)	Reports that unacceptable frame has been received
Reset (RSET)	Resets sequence numbers
Test (TEST)	Used to exchange test signals
Exchange Identification (XID)	Used to exchange identity and status

The first five commands are used to initialize, change or terminate modes and are known as **mode-setting commands**. A station receiving such a command normally acknowledges its receipt with the UA response. A change of mode causes the I-frame sequence numbers to be reset to zero. Other responses that may result from a mode-setting command are RIM, RD and DM. The UI and UP frames are used to exchange control information between stations. The response FRMR is returned when an error occurs in a received frame. It is normally followed by a RSET command, resulting in send and receive sequence numbers being reset.

7.4.3 Link management

For data to be exchanged over an HDLC link, a connection must first be set up. Normally, this is achieved by the transfer of either an SNRM or an SABM command, depending on whether normal response or asynchronous balanced mode is being established. The receive station will respond with a UA frame if it is in a position to set up the link. A typical transfer of frames illustrating the link management aspects of an ABM link is shown in the signal flow diagram of Figure 7.8.

In this mode, the settingup or clearing of a link may be initiated by either station. The link is set up by a SABM command and cleared by a DISC command. The UA response is used in both cases to acknowledge the successful acceptance of the commands.

7.4.4 Error control

This is achieved in HDLC by the use of ARQ and a cyclic error-detecting code. Prior to transmission the block of data consisting of address, control and information fields is treated as a single binary number and modulo-2 divided by a generator polynomial which is specified by the ITU-T as $x^{16} + x^{12} + x^5 + 1$

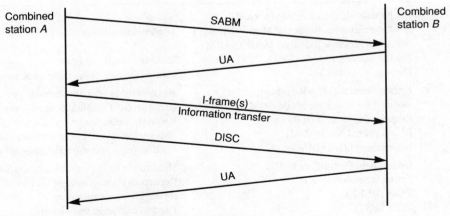

Figure 7.8 Frame transfer in an ABM link.

(10001000000100001 in binary). The remainder of this division constitutes the FCS. The flags are then added and the frame transmitted. At the receive station the received frame is stripped of the flags and divided by the same generator polynomial. The remainder of this division provides a syndrome which is zero if no errors have occurred. In the event of a nonzero syndrome, a REJ frame is returned if go-back-n ARQ is being used and a SREJ frame is returned if selective-repeat ARQ is being used.

7.4.5 Flow control

The flow control aspects of HDLC vary slightly depending on whether NRM or ABM is being used. In NRM, data flow is carried out under the control of the primary station. A typical NRM data transfer between primary and secondary stations using a go-back-n ARQ strategy is shown in Figure 7.9.

In this figure, data flow is from the primary to the secondary station only, so that the I-frames are acknowledged by supervisory frames. Each station keeps a count of the send sequence number, F(S), of the next I-frame to be sent and a count of the receive sequence number, F(R), of the next I-frame to be received and it is these counts, along with the sequence numbers inside frames, that allow the flow control to function. When a station receives a frame it compares its own receive sequence number, F(R), with the frame's send sequence number, N(S). If these two numbers are equal then the frame is accepted and if they are not equal the frame is

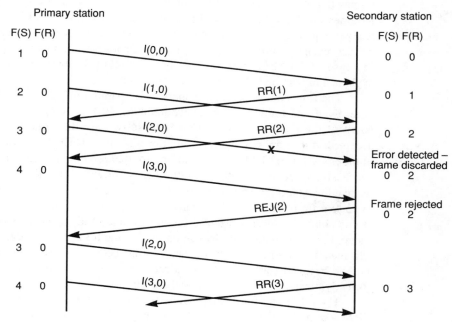

Figure 7.9 NRM data transfer.

rejected. The receive station then uses the value of its receive sequence number in the resulting acknowledgement. Thus I-frame (0,0) in this example is accepted because $F(R) = N(S) = 0$ and the frame is positively acknowledged by the supervisory frame RR(1) which has a receive sequence number of 1, indicating that one frame has been received correctly. Remember from Section 7.4.2 that supervisory frames contain only a receive sequence number whereas I-frames contain both send and receive sequence numbers. The orderly flow of I-frames from the primary is disrupted in this case by the detection at the secondary of an error in I-frame (2,0) which is immediately discarded. When I-frame (3,0) arrives at the secondary there is now a sequence mismatch between the frame, which has a send sequence number 3, and the secondary station, which has a receive count of 2 and is therefore expecting to receive a frame with a send sequence number 2. This causes I-frame (3,0) to be rejected and a negative acknowledgement of REJ(2) to be returned indicating that the primary should retransmit I-frame (2,0). The primary will now go back and transmit this frame and the subsequent I-frame (3,0) again.

In the case of the ABM, the flow control procedure is a bit more complicated because both stations can transmit I-frames independently, that is, there is full-duplex transmission. Once again the best way to understand the flow control procedure is by means of a flow diagram. A typical ABM data transfer between two combined stations using a go-back-n ARQ strategy is shown in Figure 7.10. The link employs a window mechanism which, since three bits are allocated for a frame's send and receive sequence numbers, operates with modulo-8 numbers. This means that the window size is restricted to 7, that is, the maximum number of frames that can be transmitted without receiving an acknowledgement is 7. Since I-frames are flowing in each direction and each frame contains both send and receive sequence numbers, these sequence numbers can be used to acknowledge the correct receipt of a frame rather than using separate acknowledgement frames as in the case of NRM operation. This type of acknowledgement process, as was explained in Section 7.2, is known as piggybacking.

The flow control procedure operates in a similar way to NRM. As each frame is received, the frame send sequence number, $N(S)$, is compared with the receive count of the receive station, $F(R)$. If these two are the same then the frame is accepted; if they are not the same the frame is rejected. The frame receive sequence number, $N(R)$, is then used as an acknowledgement that the frame has been successfully received at the remote end as shown. Thus, the first frame to be received by station B, station A's frame I(0,0), is acknowledged by the next I-frame to be sent from station B which carries the sequence numbers (2,1) indicating that, at this point, station B has sent 2 previous frames and has successfully received 1 frame. Likewise, the receipt by station A of frame I(3,2) from station B is acknowledgement that frame I(1,0) from station A has been correctly received at station B. This is because frame I(3,2) has a receive count equal to 2 and station B has correctly received two I-frames (I(0,0) and I(1,0)).

The procedure for dealing with erroneously received frames is the same as in NRM. Frame I(2,1) from station A in Figure 7.10 is received erroneously at station B and is discarded. The next frame transmitted by station B is frame I(4,2) indicating that only 2 frames have been received successfully. The next frame to arrive at station *B* (frame I(3,2) from station *A*) is rejected because its send sequence number of 3 does not match station *B*'s receive sequence number which has remained at 2, and the negative acknowledgement REJ(2) is sent by station *B*. The effect of full-duplex working in this example is such that, by the time the REJ(2) frame reaches station *A*, it has transmitted two further frames (I(4,3) and I(5,4)), both of which are rejected at station *B* because their frame send sequence numbers do not match the receive count at station *B*. Once station *A* receives the frame REJ(2) it retransmits its original frame I(2,1) which is now renumbered I(2,5) as a result of further frames having been successfully received by station *A*. It can be seen from this example that, for a go-back-*n* strategy to work successfully, there needs to be a manageable level of errors or there will be a loss of throughput due to the retransmission of frames. ABM allows for efficient full-duplex transmission on a point-to-point link and this mode of operation is incorporated in the link level of the ITU-T X.25 protocol for packet switched networks.

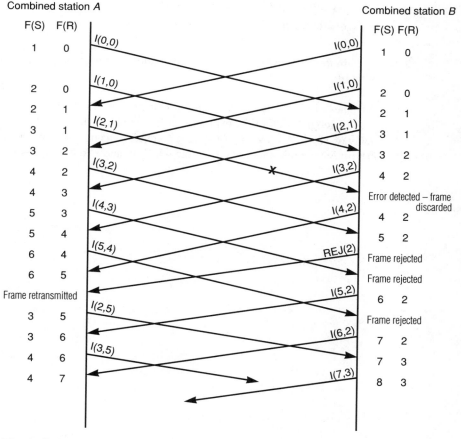

Figure 7.10　ABM data transfer using go-back-*n* ARQ.

Exercises

7.1 Two computers are connected by a 2.4 kbps link using a stop-and-wait ARQ strategy. The link has a propagation delay of 20 ms and a round-trip processing delay of 10 ms. Data is transmitted using a frame size of 896 bits of which 868 bits convey information. The acknowledgements contain 28 bits. Determine the throughput and link utilization.

7.2 A frame of data of length 1024 bits is transmitted over a link with a bit error rate of 10^{-3}. Determine the probability that a frame will be received erroneously.

7.3 Data is transmitted over a long-distance, half-duplex radio link at a rate of 7.2 kbps using a stop-and-wait ARQ strategy. Frames have a block length of 1024 bits of which 32 are non-information bits. If the propagation delay is 15 ms and processing delays and acknowledgement transmission time can be neglected, determine:

(a)   the throughput in the absence of errors,

(b)   the throughput in the presence of a bit error rate of 10^{-4}.

7.4 Explain the meaning of the term 'send window', as used in a data communication network.

7.5 A frame-oriented data communication system operates at a bit rate of 48 kbps with a frame length of 2040 bits over a long-distance link which produces a propagation delay of 18 ms. A flow control system is required using a window mechanism. Determine the minimum window size which would allow for optimum throughput.

Packet switched systems

8

This chapter describes the differences between circuit switched and message switched systems and explains why packet switched systems have been developed. Most of the chapter is concerned with the ITU-T X.25 packet switched network access protocol with particular reference to the UK's Global Network Services. The chapter concludes with a look at the more recent frame relay protocol.

8.1 Switched communication networks

Switched communication networks may be classified into the following three broad areas:

- circuit switched network
- message switched network
- packet switched network

8.1.1 Circuit switched network

The PSTN is a circuit switched network. In this type of network the user is given exclusive use of a circuit for the duration of the call. Its main characteristics are as follows:

(1) Real-time conversation can take place for as long as is desired.

(2) A finite time is required to establish a call. If it is not possible to establish a call, the network generally gives no indication when communication can be established. However, the widespread use of answer machines means that calls can be established and messages can be left in the absence of the called party. Hence instant communication is now not quite so crucial.

(3) When terminals establish communication, the channels and equipment used are exclusive to that call and are unavailable for other communications.

Circuit switching tends to be rather an inefficient means of communication. A channel is dedicated to the user for the duration of the connection, even if no information is being transmitted. For a typical voice transmission there will be lots of pauses and gaps in the conversation and the network utilization is nowhere near 100%. For a terminal-to-computer connection the channel may also, in effect, be idle for the majority of the time. Circuit switched circuits also have a long call set-up time which is totally wasted if the call cannot be established. However, its biggest advantage is that once the circuit is established it is transparent to the users. Information may be transmitted at a fixed data rate in real time with the only delay being propagation delay.

8.1.2 Message switched network

No direct channel is established between the transmitter and receiver. Instead there is a channel between each terminal and a facility whereby the transmitted message can be stored and subsequently retransmitted over the next available channel. Message switching systems are often referred to as **store-and-forward** systems. The advantage of the system is that as soon as one message has passed over a channel, another message, perhaps with a different destination, can pass over the same channel. A message may have to wait for a channel to become available. It is then stored and placed in a queue to await its turn for transmission. This system leads to high channel utilization and it is very suitable for one-way delivery of messages, but is unsuitable for interactive communication if a fast response is required.

8.1.3 Packet switched network

A derivation of message switching, which is of more interest to computer communications users, is packet switching. As in message switching, no direct channel is established between the transmitter and receiver, instead there is a channel between each terminal and a facility whereby the message the terminal transmits can be stored and subsequently retransmitted over the next available channel. The message is broken into short fixed-length segments called packets. The choice of path through the network for each packet is determined by the traffic on the network at the time and the type of packet switching system used. There are two standard methods of handling this stream of packets: datagrams and virtual circuits.

Datagrams

In a datagram service each packet is treated as a separate entity with no relationship to other packets. Each packet must contain both source and destination addresses. Suppose in Figure 8.1 that terminal A has a three-packet message to send to D. It sends out the packets in the order 1-2-3 to node 1. On receipt of each packet, node 1 must make a routing decision. Packet 1 comes in and node 1 may determine that

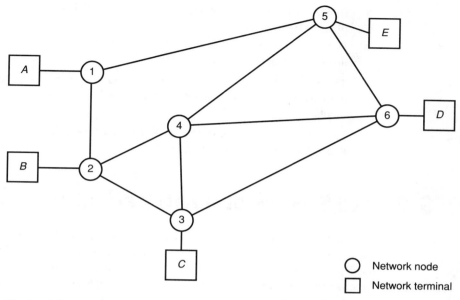

Figure 8.1 Generic switching network.

its queue for node 5 is shorter than its queue for node 2, so it queues the packet for node 5. Ditto for packet 2. But for packet 3 node 1 finds that its queue for node 2 is the shortest and so queues packet 3 for that node. Hence the packets, each with the same destination address, do not all follow the same route. It is also possible for packet 3 to reach node 6 before packets 1 and 2. It is up to station *D* to sort out how to reorder the packets. In this technique each packet is treated independently and is called a **datagram**, with obvious similarities to the telegram. It is primarily intended for the transfer of short single-packet messages.

Virtual circuits

If messages contain multiple packets, the virtual circuit/logical channel method is normally used. When using this method, prior to sending any information the source DTE sends a special call request packet to its local **Packet Switched Exchange** (PSE). The function of the call request packet is to set up a suitable route through the network. Once the route has been set up all subsequent packets from the calling DTE follow the same route to the destination DTE. As all the packets follow the same route the channel is called a **virtual circuit**. Although a virtual channel exists between the two DTEs, in that each packet travels over the same path, the packets do not have exclusive use of any individual link. Any link may have a number of packets passing along it but travelling to different destinations. Packet switching systems have the following advantages:

- Link efficiency is greater than for a circuit switched system, since a single link can be shared by many packets.

- Data rate conversions can be carried out, since each terminal connects to its node at its proper data rate.

- There is no such thing as blocking as in a circuit switched system. When the network becomes overloaded packets are still accepted but the delay time increases.

- Packets can be given priorities such that high-priority packets experience less delay than low-priority packets.

8.2 X.25 packet switched network

The main packet switching system used in the United Kingdom is British Telecom's implementation of the ITU-T X.25 standard as part of its Global Network Services (GNS). This system enables data to be sent digitally at speeds up to 48 kbps across the country. There are also links known as **gateways** to packet switched networks in most countries of the world.

X.25 is the ITU-T standard network access protocol. It defines the interface between the DTE and a PSDN. It is a set of protocols corresponding to the first three of the OSI layers. As it is an access protocol it covers the DTE to DCE interface. Therefore an X.25 network is only defined as this interface and the internal working of the network is up to the service provider; all that is required is that X.25 data entering the network also leaves it as X.25 data.

British Telecom's PSDN consists of a number of Packet Switched Exchanges (PSEs) which are linked to form a mesh as shown in Figure 8.2. Some computers are connected directly to the network but small-volume users such as e-mail users are connected via a **Packet Assembler/Disassembler** (PAD). A PAD takes asynchronous traffic and converts it into packets which can then be sent through the PSDN and vice versa. Large-volume users may have a PAD on their own premises where the volume and type of traffic demands it. Small-volume users such as e-mail users may well use a dial-up PAD.

The X.25 standard contains three levels of protocol:

- *Level 1* is equivalent to the OSI physical layer and lays down the rules necessary to establish a physical link between a terminal and the Packet Switched Network. The interface is specified by either of the ITU-T recommendation X.21 or, more likely, X.21bis.

- *Level 2* is equivalent to the OSI link layer and its function is to provide reliable transmission of data on the link between packet switched exchanges. It achieves this by transmitting the data as a sequence of frames. The link layer standard is referred to as LAP-B (link Access Protocol-Balanced) and is a subset of the HDLC described in Chapter 7.

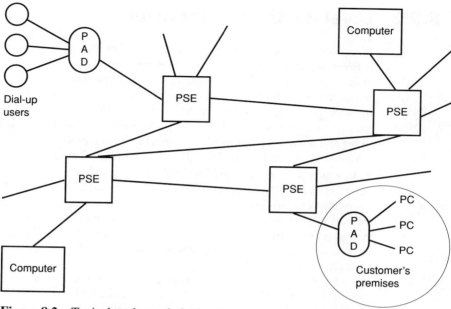

Figure 8.2 Typical packet switched network.

- *Level 3* is broadly equivalent to the OSI network layer and is the higher level
  protocol that establishes the packet formats and the control procedures
  required to set up a call and exchange information with other terminals or
  computers.

8.2.1 Level 1 – the physical level

X.21 and X.21bis define the standards used between the DTE and the DCE. The
function of the standard is to provide a full-duplex, bit-serial, synchronous transmission path between the DTE and the DCE.

X.21 bis has been defined for use with existing analogue networks (modems).
It is a subset of EIA 232D/V.24 (described in Chapter 5) hence existing equipment
can be readily interfaced using this standard plus additional software.

8.2.2 Level 2 – the link level

As has already been indicated, the function of the link level is to provide reliable
transmission of data between PSEs. The frame structure and error and flow control
procedures are based on the HDLC protocol described in Chapter 7. We shall return
to the link level in Section 8.2.4 to consider some of its functions once we have
looked at the packet level (level 3).

8.2.3   Level 3 – the packet level

Most computers tend to send data in bursts, rather than in a steady constant flow. In many cases, unlike telephone calls, it is not essential to have a constant and short propagation time. The advantages of breaking the data into packets is that other computers can send data down the same high-speed link between PSEs. GNS uses the virtual circuit technique mentioned in Section 8.1.3. A special call request packet is sent initially which carries the **Network User Address** (NUA) of both the caller and the destination in addition to a unique reference number called a **Logical Channel Number** (LCN). The LCN and the link it comes from are noted by the PSE, which then replaces the LCN with a new number and sends it forward on the outgoing link in the direction of the destination DTE. This process is repeated at every PSE until the call request packet reaches its destination DTE. Then, assuming the call is accepted, an appropriate response packet is returned to the calling DTE. At this point a virtual circuit is said to exist between the two DTEs. All subsequent packets relating to this call are assigned the same LCN references and travel over the same channel.

Each PSE now contains a routing table, an example of which is shown in Table 8.1 for the PSE of Figure 8.3. The routing table is simply a look up table. From Table 8.1 it can be seen that a packet received from link *A* with LCN 3 will be sent down link *B* with a new LCN 5 attached to it. With this method the routing intelligence is held in the PSE rather than in the message as in a datagram system. Whenever a call request packet is received by a PSE it determines the best currently available route, and changes its routing table to accommodate the new call. This circuit is called a virtual circuit as each packet with the same LCN will travel over the same route. Note that virtual circuits are also bidirectional.

Table 8.1   Typical routing table.

Input link to PSE		Output link from PSE	
Link	*LCN*	*Link*	*LCN*
A	1	B	1
A	2	C	4
A	3	B	5
A	4	D	3

The format of a packet is specified by the ITU-T X.25 recommendation. Each packet consists of two parts: the header and the data. The header consists of three octets (an octet is an 8-bit word) and the number of octets in the data field depends on the packet type. Figure 8.4 shows the format of a call request packet:

- The header, which consists of the first three octets, is common to all packets.

- Octet 1 contains the general format identifier which indicates among other things the ARQ sequence count which can be either modulo-8 or modulo-128.

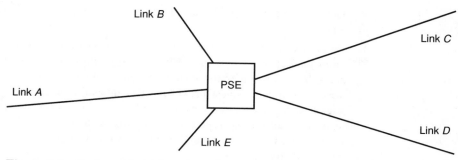

Figure 8.3 Packet switched exchange links.

Octets	Bits							
	8	7	6	5	4	3	2	1
1	General format identifier (GFI)				Logical channel group number (LCGN)			
2	Logical channel number (LCN)							
3	Packet type identifier							
	0	0	0	0	1	0	1	1
4	Calling DTE address length				Called DTE address length			
5	DTE address(es)							
					0	0	0	0
6	Facility length							
7	Facilities							
	Call user data							

Figure 8.4 Format of a call request packet.

It also contains the logical channel group number indicating the type of call. Four types of call are possible :

- Permanent virtual circuit (PVC). British Telecom calls this a Permanent Data Call.
- Switched virtual circuit (SVC). British Telecom calls this a data call of which there are three types: incoming only (SVC), both-ways SVC and outgoing only SVC.

The UK's Global Network Services provides up to eight logical channel groups for each physical X.25 connection.

Group	Type
0, 1	PVC
2, 3	SVC incoming calls only
4, 5	SVC both-way calls
6, 7	SVC outgoing calls only

- Octet 2 contains the LCN and this changes as the packet passes through each PSE.

- Octet 3 is the Packet Type Identifier (PTI), which specifies the function of each packet as shown in Table 8.2. Note that the designation changes according to whether the packet is being sent from DCE to DTE or vice versa. For a call request packet the PTI takes the value 00001011. When a packet with this PTI arrives at a PSE the information in the data field is analysed, and a route is selected. The link is then chosen and a free logical channel is assigned to the call request packet. The new LCN is in octet 2 as described above.

- Octet 4 specifies the number of digits in the address of the calling and called DTEs.

- Octet 5 specifies the calling and called DTE addresses. The address field must end on an octet boundary; this means that padding, consisting of a number of binary 0s, may have to be included at the end.

- Octet 6 (assuming only one octet is used for the calling and called DTE addresses) gives the length in octets of the facilities used.

- Octet 7 gives the facilities to be used, for example, reverse charging may be specified.

- Finally, the caller may include in the call request packet up to 16 octets of user data which, for example, could carry some user-level identification for login purposes.

Data transfer

Once the virtual circuit has been set up data can be sent. The PTI octet 3 now has bit 1 set to 0 for data transfer. The remaining part of octet 3 now has send, P(S), and receive, P(R), sequence numbers. These sequence numbers are used to regulate the flow of packets by a technique known as **flow control**. Each packet sent is numbered with a 3-bit number (modulo-8) P(S). As the system is bidirectional the receive P(R) gives an indication of the packets received. The value of P(R) is set to the number of the packet it expects to receive next. For example, if the last packet successfully received had a value of P(S) = 4 then P(R) is set to 5, that is the number of the packet it expects to receive next. So the P(R) field in the packets leaving a PSE are filled with the sequence numbers of the packets the switching exchange expects to receive next. The elegance of the system is that packets in one direction along the virtual circuit carry sequence and control information for packets in the reverse direction. This system where packets in one direction carry both data and ARQ control for the reverse direction is known as **piggybacking** (see Chapter 7).

The format of a data packet is shown in Figure 8.5. The General Format Identifier (GFI) includes a Q bit and a D bit. The Q bit is a data qualifier bit which allows two levels of data to be sent in X.25 data packets. Although this bit is reserved it is not in fact used. The D bit is delivery confirmation and is used for

Table 8.2   X.25 packet type identifier.

Packet type		Octet 3 bits							
From DCE to DTE	*From DTE to DCE*	8	7	6	5	4	3	2	1
	Call set-up and clearing								
Incoming call	Call request	0	0	0	0	1	0	1	1
Call connected	Call accepted	0	0	0	0	1	1	1	1
Clear indication	Clear request	0	0	0	1	0	0	1	1
DCE clear confirmation	DTE clear confirmation	0	0	0	1	0	1	1	1
	Data and interrupt								
DCE data	DTE data	X	X	X	X	X	X	X	0
DCE interrupt	DTE interrupt	0	0	1	0	0	0	1	1
DCE interrupt confirmation	DTE interrupt confirmation	0	0	1	0	0	1	1	1
	Flow control and reset								
DCE RR (modulo-8)	DTE RR (modulo-8)	X	X	X	0	0	0	0	1
DCE RR (modulo-128)[1]	DTE RR (modulo-128)[1]	0	0	0	0	0	0	0	1
DCE RNR (modulo-8)	DTE RNR (modulo-8)	X	X	X	0	0	1	0	1
DCE RR (modulo-128)[1]	DTE RR (modulo-128)[1]	0	0	0	0	0	1	0	1
	DTE REJ (modulo-8)	X	X	X	0	1	0	0	1
	DTE REJ (modulo-128)[1]	0	0	0	0	1	0	0	1
Reset indication	Reset request	0	0	0	1	1	0	1	1
DCE reset confirmation	DTE reset confirmation	0	0	0	1	1	1	1	1
	Restart								
Restart indication	Restart request	1	1	1	1	1	0	1	1
DCE restart confirmation	DTE restart conformation	1	1	1	1	1	1	1	1
	Diagnostic								
Diagnostic[1]		1	1	1	1	0	0	0	1
	Registration[1]								
	Registration request	1	1	1	1	0	0	1	1
Registration confirmation		1	1	1	1	0	1	1	1

[1] Not necessarily available on every network.
Note: A bit which is indicated as 'X' may be set either to 0 or 1.

Octets	8	7	6	5	4	3	2	1
1		General format identifier			Logical channel group number			
	Q	D	0	1				
2				Logical channel number				
3		P(R)			M		P(S)	0
				User data				

Figure 8.5   Format of a data packet.

Octets	8	7	6	5	4	3	2	1
1		General format identifier			Logical channel group number			
	0	0	0	1				
2	Logical channel number							
3	P(R)				0	0	0	1

Figure 8.6 Format of an RR packet.

end-to-end significance between the souce and destination. A DTE can set the D bit in data packets if it wishes the returned P(R) values in received data packets to have end-to-end significance, in which case the DTE can be 'certain' that the data packets thus acknowledged have reached the destination. If the D bit is not set, the local DCE alone decides on the P(R) values to return. In octet 3 an additional feature is the **more** (M) **bit** which may be set to indicate to the receiving DTE that the data is longer than can be fitted into a single X.25 packet. Hence in, say, a three-packet message the M bit would be set in the first and second packets. While the M bit is preserved in the X.25 network, it is not used by X.25 in any way and is used only for end-to-end significance.

If data is being sent in only one direction from, say, PSE *A* to PSE *B* then PSE *B* will send a return flow control packet of the type Receiver Ready (RR), Receiver Not Ready (RNR) or Reject (REJ). The format of an RR packet is shown in Figure 8.6.

8.2.4 Level 2 – the link level

The packet level is concerned with setting up virtual circuits from one end of the connection to the other, that is, between DTEs. The link level is concerned with transporting packets from one PSE to the next, with an acceptably low error rate. It is not concerned with logical channels, call set-up packets and so on. It just receives the packets and transports them to the other end of the link. When packets are sent down a link there must be a means of indicating when one packet finishes and the next one starts. This is done by sending each packet in a frame and each frame starts with a flag. The format for a LAP-B I-frame is shown in Figure 8.7.

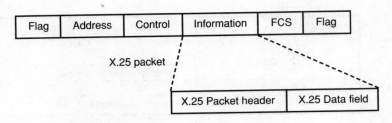

Figure 8.7 Lap-B I-frame.

Table 8.3 Examples of bit stuffing.

Type	Natural Form	Bit stuffed
Flag	01111110	01111110
Data	01111101	011111001
Data	11111100	111110100

The flag has a unique pattern which enables the ends of the frame to be identified. The flag is defined as 01111110. As there is a possibility that this pattern may occur naturally in a data stream, bit stuffing is used to prevent its occurrence (see Section 7.4.1). The sending end ensures that it never sends a string of six binary 1s unless it is a flag. If six binary 1s appear in any of the other fields then a binary 0 is inserted after the fifth bit. At the receiver a string of five consecutive binary 1s is recognized and the sixth bit binary 0 is automatically removed (destuffed). Examples of bit stuffing are shown in Table 8.3.

In a similar manner to the packet level there are different frame types in the link level; both levels are concerned with moving information between points and use similar methods to do so. The address field identifies the destination. There are three frame types: information, supervisory and unnumbered. These frames are identified by the control field which is the second octet. The information and supervisory frames are concerned with error and flow control and use a similar principle to the go-back-n ARQ protocol used at the packet level. Both these frames carry sequence numbers N(S) and N(R) to implement the go-back-n ARQ protocol. The unnumbered frames are used to set up and clear down the link and do not contain sequence numbers. However, as most links are in operation continuously, the frame level set up and clear is only rarely used. The control field format is shown in Figure 8.8.

Figure 8.8 shows that the I-frame is identified by bit 1 being set to binary 0. Bits 2, 3, 4 and 6, 7, 8 are used for sequence numbers N(S) and N(R), respectively. This is very similar to the data PTI of the packet level. Again, as in the packet level, there are three supervisory frames called Receiver Ready (RR), Receiver Not Ready (RNR) and Reject (REJ). The use of these frames is similar to the packet level but they are dealing with link operation rather than end-to-end operation. They are used when there is no data flowing in the reverse direction and flow control information needs to be sent:

- An RNR frame is sent to indicate an acknowledgement of frames up to and including frame N(R) – 1 together with an indication that the receiver cannot accept any more. The maximum number of unacknowledged frames that can be in transit at any time is the window size. When the window size is reached RNR frames are sent to stop transmission of further frames. The normal window size is 8 but an extended format allows a window size of 128.

- An RR frame is used to acknowledge positively an I-frame when there is no I-frame to send in the reverse direction. An RR frame may also be sent to clear an existing RNR state.

Format		Command		Response	Encoding 1 2 3 4 5 6 7 8							
Information transfer	I	Information			0	N(S)			P	N(R)		
Supervisory	RR	(receive ready)	RR	(receive ready)	1	0	0	0	P/F	N(R)		
	RNR	(receive not ready)	RNR	(receive not ready)	1	0	1	0	P/F	N(R)		
	REJ	(reject)	REJ	(reject)	1	0	0	1	P/F	N(R)		
Unnumbered	SABM	(set asynchronous balanced mode)			1	1	1	1	P	1	0	0
	DISC	(disconnect)			1	1	0	0	P	0	1	0
			DM	(disconnect mode)	1	1	1	1	F	0	0	0
			UA	(unnumbered acknowledge-ment)	1	1	0	0	F	1	1	0
			FRMR	(frame reject)	1	1	1	0	F	0	0	1

Figure 8.8 Control field format.

- An REJ frame is used to inform the sender that a frame has been received which contains an error, and that it needs to retransmit the frame with the sequence number N(R).

There are five unnumbered frames which carry out the following functions:

- The set asynchronous balanced mode (SABM) is used to activate the link when it has not been previously in operation.

- If the receiving end is able to accept activation of the link it replies with an unnumbered acknowledgement (UA).

- If the link needs to be made non-operational it is deactivated by sending a disconnect frame (DISC).

- Once the link is deactivated, disconnect mode (DM) frames are sent to indicate that the link is logically disconnected.

- The FRMR frame is concerned with an extended format operation. LAP-B has an extended format mode which has 7-bit sequence numbers and allows a window size of 128. The FRMR frame is sent by a receiver unable to comply with a request to set up an extended format link.

8.2.5 Packet assembler/disassembler

The prime function of a PAD is to connect asynchronous terminals to a PSDN. The PAD must therefore perform all the X.25 protocol functions on behalf of the terminal, its aim being to make the packet switching network transparent to the user.

There are a number of ITU-T standards that define the operation of the PAD and the associated character-oriented terminal. ITU-T recommendation X.3 defines the basic operation of the PAD. Recommendation X.28 defines the interface requirements between the terminal and the PAD. Finally, X.29 defines the interface between the packet mode terminal and the PAD.

A very common mode of operation between computer and terminal is called **echoplex**. In this mode a character transmitted from the terminal is not displayed on the screen until it has made the round trip to and from the host computer. This causes problems for the packet switching network as follows:

- The round-trip time delay may be relatively long.

- It may be expensive as most PTTs charge on volume-oriented traffic and if the character transits the network twice the user is charged more for it.

- It generates more traffic which increases loading and delay in the network.

A function of the PAD is therefore to provide local echoing of characters rather than requiring the characters to be echoed end-to-end by the host.

The cheapest way of operating a packet switching system is to ensure that every packet is full. If a single character is sent the overhead is three octets, which has more bits in it than the data itself. The PAD therefore needs to know the rules which allow it to send a packet. A common rule in e-mail systems is that the PAD stores the characters but does not send a packet until it receives a carriage return (CR) character.

To facilitate the use of a PAD, all the parameters associated with the terminal have default values, so that only those parameters whose values differ from these need to be changed.

8.3 Frame relay

Packet switching systems have been in operation for over 25 years and at the time of their development the BERs were rather high, typically 10^{-3} for a modem link across a voice-grade circuit, making error correction essential. The system used in X.25 can achieve a thousandfold improvement on this figure. To achieve this there is a considerable overhead built into packet switching schemes to compensate for the errors. The overhead includes additional bits added for error checking and for processing at end and intermediate nodes to detect and recover from errors. As a result of this overhead and the ARQ systems used the rate of throughput is quite low. The original packet switching networks were designed with a data rate to the end user of only 48 kbps.

With modern transmission systems the high overheads incorporated in X.25 systems are unnecessary. A modern digital trunk line sourced from an optical fibre network has a BER of around 10^{-8}. At this sort of BER the link-level error

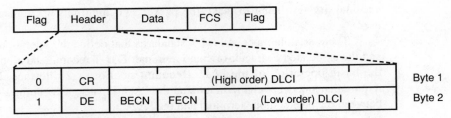

Figure 8.9 Frame relay frame format.

correction of X.25 is unnecessary. LAN traffic commonly operates at speeds of around 10 Mbps. To avoid degradation in performance WANs need to operate at comparable speeds. Frame relay networks are currently operating at speeds of 2 Mbps, and speeds of 34 Mbps are not far away. The method of achieving these higher data rates is to remove all the unnecessary overheads associated with error control. Frame relay, unlike X.25, detects frames that contain errors but performs no error recovery – frames that contain an error are discarded. Frame relay assumes that the higher OSI layers will cope with such losses.

In essence, X.25 is a 3-layer protocol working at layers 1, 2 and 3 of the OSI. Frame relay effectively removes layer 3 and is restricted to layers 1 and 2. It uses a frame structure which has some resemblance to that of X.25. The frame contains flags for marking the beginning and end of a frame, a header field, an information field and a frame check sequence. A frame format is shown in Figure 8.9. In X.25 the routing information is contained in level 3, the packet level, whereas in frame relay it is contained in level 2, the frame level. The routing formation is contained in the frame header as a Data Link Connection Identifier (DLCI). This is the equivalent of X.25's virtual circuit number.

As there is no layer 3 processing, frames proceed across nodes with little delay. No acknowledgement signals are used; if the FCS indicates an error the frame is simply discarded. As there is no acknowledgement, signal flow control is operated differently from X.25. The frame header contains two bits which may be set by a frame handler that detects congestion. Once the bits are set a frame handler should not clear them before forwarding a frame. The bits constitute signals from the network to the end user. The two bits are:

- **Forward Explicit Congestion Notification** (FECN)   When this bit is set it informs the receiving device that there is congestion in the direction the frame has been travelling.

- **Backward Explicit Congestion Notification** (BECN)   When this bit is set it informs the receiving device that there is congestion in the opposite direction to which the frame has been travelling.

The user response is as follows:

- On receipt of a BECN signal the user simply reduces the rate at which frames are transmitted until the signal ceases.

- Receipt of a FECN signal requires notification of the peer user to reduce its flow of frames although frame relay does not support this notification. Hence this control must be done at a higher layer, such as the transport layer.

A further feature of frame relay is the indicator **Discard Eligibility** (DE) which, when set, indicates that the frame should be discarded in preference to other frames in which it is not set. The DE facility makes it possible for the user to send more frames than it is allowed on average. In this situation, the user sets the DE indicator on the excess frames and the network forwards these frames only if it has the capacity to do so.

Frame relay's principal advantage is its speed of operation and, although transmission errors may still occur, modern LAN terminals invariably operate their own level 3 protocol providing recovery for the occasional layer 2 transmission failure.

Exercises

8.1 Explain, giving the relative advantages and disadvantages, what is meant by circuit switched and packet switched networks.

8.2 The following data stream has been received at a packet switched exchange. Destuff the stream and clearly indicate any flags:

010011111001110011111101111100011110000

8.3 How would a sequence received before destuffing which consists of seven consecutive binary 1s be interpreted?

8.4 Briefly describe the principal functions carried out in each of the three levels (physical, link, packet) in the X.25 system.

8.5 Describe how a packet switched call can be set up across a network and explain the following terms: routing table, logical channel numbers and virtual circuit.

8.6 Briefly describe the format of an X.25 link-level frame, naming its principal elements.

Local area networks 9

LANs encompass a small physical area of no more than a few kilometres in radius and are usually confined within one site. The number of stations may be as many as several hundred. LANs use relatively high signalling rates typically between 1 and 100 Mbps. Messages are transmitted as a series of frames or packets, which may not necessarily be of fixed length, using transmission media having relatively low error rates. Frame propagation delay tends to be small, unlike WANs. Such frame-based transmissions are principally suited to data rather than applications such as voice and video. This is because delay may be unpredictable and hence, on occasion, excessive for real-time use and transmission capacity is generally insufficient to meet the demands of video.

A LAN comprises three hardware elements, namely a transmission medium, a controlling mechanism or protocol to govern access to the transmission medium, and an interface between station and transmission medium. A software element is required to implement the protocol for intercommunication between stations.

This first part of this chapter is a review of the mechanisms to control station access to the network and makes some performance comparisons. The place of repeaters, bridges and routers within LANs is briefly discussed which enables the next section, a detailed examination of the three IEEE standard LAN types, to be fully appreciated. There then follows an explanation of a high-speed LAN in the form of fibre distributed data interface network followed by the IEEE 802.6 metropolitan area network used for the interconnection of LANs across a larger physical area, such as a city.

9.1 Medium access control techniques

The media used in LANs generally convey frames from only one station at a time although the media themselves are generally shared by a number of stations. In order to overcome the difficulties which may arise through sharing, a **Medium Access Control** (MAC) mechanism is necessary. A MAC protocol merely regulates access to the medium in an orderly fashion for correct operation and also attempts to ensure that stations obtain a fair share of its use.

First consider interconnection of stations within a LAN by the star topology (see Figure 2.13). Here a central hub station (or controller), using a suitable protocol, may control communication between stations in an orderly manner. Each station has its own designated communication channel to the hub. The channels are not shared, as is the case with most LANs, and hence there is no contention when two, or more, stations try to access the medium simultaneously. Such a network does not require a MAC protocol.

However, networks which employ a bus or ring topology have, in essence, only a single medium over which all messages are transmitted. If the medium is not being used, two or more stations may simultaneously attempt an access, leading to a collision. A MAC technique is therefore required to regulate access by stations to the medium to reduce, or eliminate, such collisions. There is also the danger that once a pair of stations have established communication, all other stations may be excluded, perhaps indefinitely, or at least for a considerable period of time.

Bus and ring topologies do not necessarily require any separate network control function for operation or the detection and control of, and recovery from, abnormal network conditions. Rather, each station can be equally responsible, hence control is said to be fully distributed. In contrast, a star configuration, almost by definition, dictates that the hub acts as a central controller. Such a system is known as a **centralized control** system where stations are **slaves** to a single controlling station, often called a **master**. There is therefore no concept of distributed control within a star topology.

Three general MAC techniques exist for use within fully distributed networks:

(1) *Contention*   Here there is no regulating mechanism directly to govern stations attempting to access a medium. Rather, two or more stations may contend for the medium and any multiple accesses are resolved as they arise.

(2) *Token passing*   A single token exists within the network and is passed between stations in turn. Only a station holding the token may use the medium for transmission. This eliminates multiple simultaneous accesses of the medium with the attendant risk of collision.

(3) *Slotted and register insertion rings*   Similar in principle to token passing, but a unique time interval is granted to a station.

9.1.1 Aloha

Aloha is an example of a primitive LAN which is packet based and uses radio as a transmission medium. Such a LAN was installed by the University of Hawaii in the early 1970s to interconnect stations spread across a group of islands. Although it is an obsolete system, Aloha is an extremely useful example of a LAN for the introduction of some basic LAN concepts, which is helpful in examining the later sections of this chapter.

Each station has equal status (distributed control) and follows a strict protocol for gaining access to the medium. Only a single radio channel exists for access hence a contention MAC protocol is employed. A station wishing to transmit does

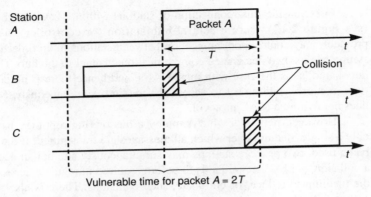

Figure 9.1 Vulnerable time – Aloha.

so immediately without checking to see if the medium is already in use. As a result, collisions frequently occur when two or more stations access the medium simultaneously. Any packets which collide are generally unusable and may be regarded as lost. A flow control protocol is employed to identify any lost packets and arrange for their repeat transmission.

We have already established one way in which a collision may occur, that is simultaneous transmission of two or more packets. Now suppose that packets are of equal, and fixed, duration T. With reference to Figure 9.1, suppose station A transmits a packet. A collision may also occur if another station commences transmission of a packet up to T seconds before Station A's packet begins. The time interval during which two or more transmitted packets may collide is known as the **vulnerable time**. Packet A is vulnerable to collision for a time equal to twice the length of a single packet, or $2T$.

9.1.2 Slotted Aloha

To reduce collisions, all stations may be synchronized in time. Packets, still of fixed duration, can only be transmitted during a time slot as shown in Figure 9.2. Although packets may still collide when two or more stations attempt to transmit in the same slot time, the vulnerable time in **slotted Aloha** is now only equal to *one* time slot, and not two as in pure Aloha. Since the vulnerable time is halved, the probability of collisions falls and throughput is improved.

Figure 9.3 compares throughput of Aloha and slotted Aloha. It is assumed that propagation delay is negligible. Throughput S is often normalized with respect to the maximum possible throughput which is equal to the bus capacity or its signalling rate. When S is normalized it may take on values only between 0 and 1. Alternatively, as is assumed here, where packet time is constant S may be regarded as the average number of packets transmitted in one packet time. G is a measure of the demand, or load, presented to the network. In Figure 9.3 we may regard G as the number of packet attempts per packet time. Clearly, G may exceed 1.

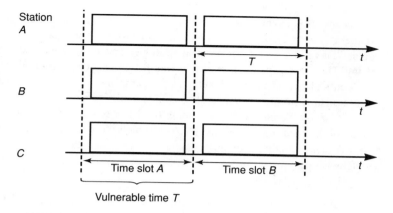

Figure 9.2 Vulnerable time – slotted Aloha.

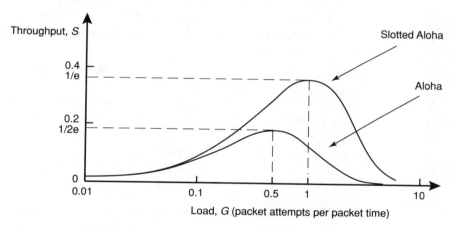

Figure 9.3 Comparison of Aloha and slotted Aloha.

An interesting feature of both curves shown in Figure 9.3 is that once the number of packet attempts per packet time, or slot, approaches 1, throughput actually reduces. This is an overload condition and is caused by an excessive number of collisions which reduces available medium time to pass packets successfully. This is further exacerbated by each collision giving rise to a retry, increasing the number of packet attempts.

9.1.3 Carrier sense multiple access

Radio LANs were developed further to increase throughput. When a station wished to transmit it 'listened' to activity, for example the presence of an RF carrier. If no carrier was present the medium was assumed to be idle and the station then made an access. This **carrier sense** strategy is also known as **Listen Before Talk** (LBT). If there is activity, a station defers its transmission for a period of time and tries again. There is a variety of deferral strategies and most make use of a randomized delay. The reason for randomizing is that if two or more colliding stations defer for an identical period they may retry at the same instant and, if the medium is free, both transmit simultaneously and so collide again.

Carrier Sense Multiple Access (CSMA) is not confined to networks using radio as the medium. Other early LANs which were cable based also made use of CSMA. The remainder of this chapter assumes that cable is the medium used for transmission.

LBT gives rise to two problems:

(1) Two or more stations may listen to the bus at the same time, both detect absence of carrier and transmit simultaneously and therefore collide.

(2) Even if only one station makes an access then, because of the finite velocity of propagation of the radio signal, a second distant station may perceive the channel to be quiet for a very small interval of time after the first station commences transmission. If the second station transmits within this interval a collision will occur.

In the extreme case stations may be located at either end of the medium. When one station sends a frame the other is unaware of its presence until the frame has propagated the full length of the medium. Up until that moment the distant station may send a frame. The vulnerable time is therefore equal to the largest time interval between any two stations in the network. No collisions will occur after this interval. In consequence CSMA is not constrained to fixed-length frames.

Collisions are greatly reduced with CSMA compared with Aloha since the vulnerable time is reduced, which in general leads to a greater throughput. There is, however, a danger in CSMA that two, or more, deferring stations may guarantee collision once the medium becomes free. The next section discusses various deferral strategies which, when used in an optimal manner, arrange for CSMA to have improved performance in terms of throughput compared with Aloha.

9.1.4 Persistence algorithms

There are a number of deferral strategies, or algorithms, used in CSMA for dealing with a station which discovers the medium to be busy.

1-persistent

A station wishing to transmit waits until the medium is free. When the medium is, or becomes, free any waiting station/s may transmit immediately. Waiting stations therefore have a probability of transmitting of 1 associated with them once the medium becomes free. This persistence algorithm means that where two or more stations are waiting, or deferring, a collision is guaranteed.

Nonpersistent

Here, if the medium is busy, a station waits a period of time determined by a probability distribution and then re-examines the medium to see if it has become free. This deferral period is also called the **back-off time**. Two or more deferring stations are unlikely to be assigned the same back-off time. Hence the probability of two or more stations retrying simultaneously, with a resultant collision, is small. In general, such an approach means that when the medium becomes free, it often may not be used immediately because stations waiting to transmit may still be deferring.

P-persistent

This algorithm is a compromise between the other two. It attempts to reduce collisions and also the time during which the medium is idle when stations are deferring. It generally offers improved performance.

Each station, upon detecting that the medium is free, transmits with a probability P. This, in comparison with the 1-persistent algorithm, reduces collisions due to two or more stations waiting for the medium to become idle. If the station fails to transmit, the probability of this occurring being $1 - P$, it defers. Usually, deferral is for one time slot in a synchronous system, or the maximum propagation delay in an asynchronous system. If, after deferring, the medium is found to be busy again, the whole procedure is repeated. A number of probabilities are used in practice. In general, the lower the probability P, the lower the probability of collision and the greater the throughput attainable. Figure 9.4 compares the various CSMA protocols along with those for Aloha. Note that in the case of P-persistence algorithms, curves for $P = 0.5$, 0.1 and 0.01 are shown.

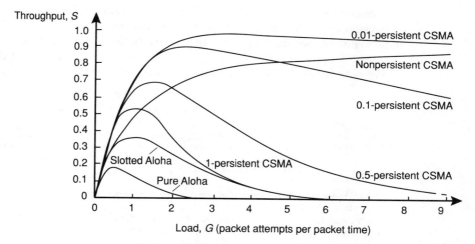

Figure 9.4 Comparison of Aloha and CSMA. Source: Stallings, William, *Data and Computer Communications*, 3rd edn, © 1991, p.350. Reprinted by permission of Prentice-Hall, Upper Saddle River, New Jersey.

9.1.5 Carrier sense multiple access – collision detect

With CSMA, any collisions which occur are not immediately apparent to a station and the whole frame is transmitted. **Carrier Sense Multiple Access – Collision Detect** (CSMA/CD) MAC uses one of the CSMA strategies above. However, with CSMA/CD, once a station commences transmission it monitors its own transmissions on the medium and if it detects a collision, ceases to transmit the remainder of its frame. Hence CSMA/CD releases the medium much earlier than is the case with Aloha and CSMA. Upon detection of a collision a **jamming signal** is transmitted immediately upon cessation of frame transmission. This ensures that all stations know that a collision has occurred; this is necessary because two colliding stations which are very close may both cease transmitting before either of their transmissions has fully propagated throughout the medium to other stations. Finally, stations perform a random back-off and then re-attempt transmission using LBT. The random back-off is achieved using one of the persistence algorithms already discussed.

The CSMA/CD algorithm may be summarized as follows:

(1)   LBT.

(2)   If free, transmit and monitor transmission.

(3)   If busy, defer.

(4)   If a collision occurs during transmission, stop transmitting.

(5)   Send a jamming signal.

(6)   Random back-off.

(7)   Retry with LBT.

Consider the case shown in Figure 9.5 where station A is located at one end of the medium and station B at the opposite. Suppose that station A sends a frame. Until the frame reaches station B, it too may send a frame. Suppose that station B chooses to send a frame at the exact moment that station A's frame arrives at B. Station A will only become aware of a collision when station B's frame has propagated the

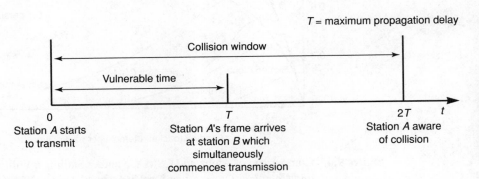

Figure 9.5   CSMA/CD collision mechanism.

full length of the medium. The **collision window** is the period during which a station's frame may be transmitted before it is aware of a collision. This equals twice the propagation delay between stations *A* and *B* and is equal to twice the vulnerable time. The jamming signal must therefore be at least equal in time to the collision window in order to ensure correct operation of the deferral mechanism used by each station.

LANs usually have frames of much greater duration than the maximum propagation time within the medium. In the worst case with CSMA/CD, the amount of time wasted in a collision equals the collision window interval plus the additional duration until the end of the jamming signal. This is appreciably less time than with CSMA where a complete frame is sent. It follows that the use of collision detection considerably reduces time wasted on the medium and so enables improved throughput compared with CSMA.

In Section 9.2 LAN performance will be examined in detail. It will be seen that CSMA/CD throughput is similar to that of CSMA for networks which comprise a small number of stations. However, where more stations are connected to a network, CSMA/CD offers higher throughput.

9.1.6 **Control tokens**

Token-based access strategies, Figure 9.6(a), rely upon the establishment of a physical or logical ring. Upon initialization of the network a single token is created and circulates around each station on the ring in turn. A station wishing to transmit a message waits for the token to arrive at its node and then commences transmission. When a station has finished transmitting, or if it is timed out, the token is released to circulate around each station in turn until seized by a station again.

The Aloha, CSMA and CSMA/CD MACs are examples of bus-based networks. In these cases the bus may be *passive* where each attached device is tapped onto the bus and does not necessarily form part of the transmission path. Stations using token-based MAC's can also be passive. However, in general, they are usually *active*, as indicated in Figure 9.6, where stations are inserted into the transmission medium and electronically relay, or repeat, incoming signals to their outgoing port.

Tokens introduce a high degree of control and order with regard to stations accessing the bus. The use of tokens eliminates collisions and hence makes for much higher throughput. Great care is needed to guard against a station which has obtained a token preventing any other stations transmitting by failing to release the token. Timers may be used to limit this. In addition, provision must be included to enable stations to be added to or removed from the network. The possibility also exists that the token may be lost or duplicated. In a well-designed token network, the MAC protocol needs to be quite sophisticated compared with contention-based protocols.

Another consideration is that of release of a token. A token may be released at the receiver (RAR) or the transmitter (RAT). The former technique has the advantage

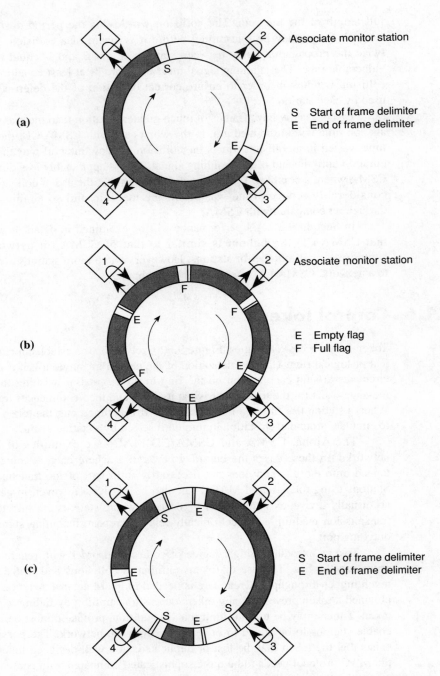

Figure 9.6 Token, slotted ring and register insertion operation: (a) token passing; (b) slotted ring; (c) register insertion.

that the medium is freed much earlier but an acknowledgement is required to confirm that the frame has been correctly received. RAT can arrange for the recipient of a frame to mark the token to indicate that it has been successfully received. This may then be checked by the transmitter. The process eliminates any need for acknowledgement and is therefore simpler to implement.

Finally, one station may be designated a **monitor**, which has responsibility for functions such as ring management and operation. An example is the creation of a token upon initialization. Another function of a monitor is to guard against the endless circulation of a frame. Suppose a station seizes a token and then transmits a frame. If this frame is never removed by any station and in consequence a new token is not released, no other station is able to send any frames.

9.1.7 Slotted ring

Figure 9.6(b) shows an active ring which contains an integral number, 6 in this case, of circulating miniframes, or slots, each of equal length. There are typically between 3 and 15 such frames, each of around 40 bits. One station acts as a monitor and is responsible for creating the frame structure. Each station is active and therefore contributes some bits between its input and output which form part of the slotted structure. Similarly, the medium has a number of bits in transit at any one time. To ensure that an integral number of frames may be accommodated within the ring, the monitor pads out the ring with additional bits, as required.

When a station wishes to transmit, it examines a Full/Empty (F/E) bit, or flag, within each received slot. If the frame is marked free the station inserts its own data in the frame and sets the F/E flag to full. When the frame returns to the sender it changes the F/E flag to empty to free the frame for other users. Restriction may be made to prevent one station from using consecutive frames to ensure fair share operation. As a precaution, the monitor also sets a monitor flag as a full frame passes and which the transmitter must reset. If a frame returns to the monitor with the monitor flag still set, the frame has not been released by the transmitter. The monitor then resets both F/E and monitor flags to prevent endless circulation of unusable frames.

9.1.8 Register insertion

Figures 9.6(c) and 9.7 illustrate the concept of **register insertion**. Each station may have only one frame in the ring at once. The start and end of each frame is indicated in some way. A station wishing to transmit places its frame in a register. It then waits for the start of the next incoming frame and instead of simply relaying the frame, it transmits its own frame and buffers the received frame into the register immediately following the bits of its own frame. Finally, when the station's own frame is sent, the station transmits the previously stored incoming frame.

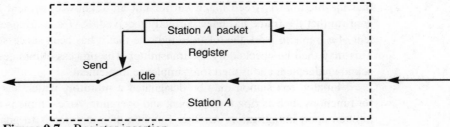

Figure 9.7   Register insertion.

Receiving stations do not remove frames but may change them to indicate an acknowledgment. It is the responsibility of the transmitting station to remove frames, but should this fail to occur frames may circulate forever. This is because, unlike slotted and token rings, there is no monitor. Since frames are inserted in the network only as required, register insertion type rings potentially offer very high utilizations. Although register insertion was used in some early LANs, token passing and slotted ring MACs tend to dominate today.

9.2 Local area network performance

In Section 9.1 we discussed the principal MAC protocols used within a LAN. As already indicated, LAN performance may vary according to the MAC protocol used, as well as other factors, such as transmission rate and number of stations. In this section we shall compare the performance of the two principal types of LAN found in practice, namely CSMA/CD and token passing. This comparison will conveniently conclude our general discussion on LANs and prepare the way for consideration of the specific details appertaining to LANs which conform to the IEEE standards.

First, we shall make a generalized analysis of both contention and token-type networks to enable performance comparison to be made between CSMA/CD and token-based networks. In assessing network performance a number of factors may be regarded as independent of the attached devices and as purely attributable to network implementation. The following factors may be thought of as variables which affect performance:

- medium transmission speed, or capacity
- propagation delay
- number of bits per frame
- medium access protocol
- load offered by stations
- number of stations
- error rate

The first two variables have a profound effect upon performance in terms of maximum throughput. They characterize the network but, from a design consideration, can be thought of as constants.

The number of bits per frame and MAC protocol are very much under the control, or at least selection, of a network designer. Load and number of stations are independent variables. This means that a system could be proposed and its performance analysed for varying load and number of stations to seek either the best network for given variables, or a limit to the variables. Errors may be controlled by overlaying protection or correction at the data link layer, or by employing retransmissions techniques at the session layer. In the absence of errors such protection inevitably leads to some reduction in throughput.

9.2.1 Measures of performance

First, we must consider basic measures of performance. Two measures of performance commonly used are as follows:

- S, the throughput of the network. This was defined in Chapter 7. An alternative way of expressing S is as the effective (or mean) data transmission rate between stations. This often includes overheads in assembling frames such as headers but does not include retries. In consequence the effective data rate is lower when considering data throughput in practice.

- D, delay – the time that elapses between a frame being assembled ready for transmission and the completion of successful transmission. It is caused by factors such as LBT, persistence algorithms and collisions with their attendant retransmissions.

EXAMPLE 9.1

Over a period of 1.8 s, 300 frames are transmitted over a 10 Mbps bus. Determine the effective throughput as a percentage of bus capacity. Assume that the average frame length is 782 bytes of which 750 convey data.

Throughput S in bps is obtained as follows:

$$\frac{\text{Number of frames} \times \text{average number of bits per frame}}{\text{transmission time}} = \frac{300 \times 782 \times 8 \text{ bits}}{1.8 \text{ s}}$$

$$= 104\,267 \text{ bps}$$

Normalizing S with respect to the bus capacity by expressing throughput against bus capacity gives:

$$S = \frac{104\,267}{10 \times 10^6} = 0.1043$$

or approximately 10% of the potential medium capacity.

If the frame overhead bits are ignored, effective throughput in terms of actual data can be determined:

$$S = 0.1043 \times 750/786$$
$$= 0.0995$$

Frame overhead represents about 4.6% of the actual throughput.

A graph of throughput S may be plotted against the load (G) presented to a network. Clearly, under low utilization, we would expect S and G to be the same. However, S is ultimately limited, mainly by the medium capacity, but also by the MAC protocol used. In consequence, as a network becomes more heavily loaded, a point occurs where any further rise in G causes no further increase in S, as shown in Figure 9.8. G includes repeat frames and any other nondata frames such as tokens.

Intuitively, we can expect delay D to increase with G. As load increases the mean time to gain access to the medium, or delay, rises. This is due to an increased probability of a collision and attendant delay before retrying in contention-based networks. In token-based networks, the busier they become the longer a given station may have to wait for all stations ahead of it to gain a token, send their frame/s and release the token. Hence mean token waiting time, or delay, is increased.

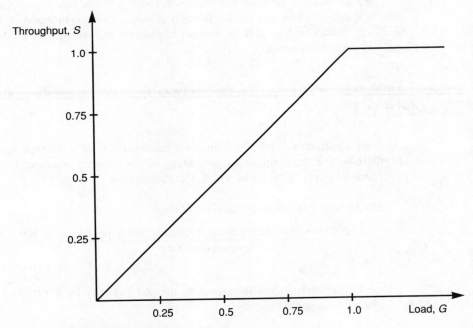

Figure 9.8 Medium utilization (ideal).

9.2.2 Propagation delay and frame transmission time

Another parameter which has a profound effect upon network performance is the ratio of propagation delay to that of frame size. This ratio is known as the **transmission coefficient**, a:

$$a = \frac{\text{propagation delay}}{\text{frame transmission time}} \qquad (9.1)$$

Propagation delay is simply the time it takes for a signal to propagate from a send station to a receiver. Frame transmission time is the time it takes to transmit all of the bits of a frame onto the physical medium. As we shall see shortly, a is a useful parameter for analysing and comparing LANs.

Alternatively, and often more conveniently, a can be expressed as:

$$a = \frac{\text{length of medium}}{\text{length of frame}} \qquad (9.2)$$

where frame transmission time is replaced by the equivalent physical distance occupied by the frame at one instant. In analysing LAN performance it is assumed that the propagation delay is a maximum. This corresponds to the medium length being the maximum, or end-to-end, length.

Consider a perfectly efficient access mechanism that allows only one transmission at a time (no collisions). Additionally, as soon as one transmission finishes, the next commences immediately and frames contain data only, with no overhead. This idealistic model enables an upper bound on performance to be determined for varying medium lengths and hence different values of a.

To determine an upper bound for S as a function of a, some other terms must be introduced:

R, data rate of channel (bps)
d, maximum distance between any pair of stations,
V, velocity of propagation (m/s)
L, frame length (assume an average if variable) (bits)

Throughput S may be expressed as:

$$S = \frac{\text{duration of one frame}}{\text{time occupied by medium for transmission of one frame}} \qquad (9.3)$$

$$= \frac{L/R}{d/V + L/R} \qquad (9.4)$$

Similarly, a may be related to R, d, v and L thus:

$$a = \frac{\text{propagation time}}{\text{frame duration}} \qquad (9.5)$$

where propagation time is assumed to be the maximum possible. Therefore:

$$a = \frac{d/V}{L/R} \qquad (9.6)$$

$$= \frac{Rd}{LV} \qquad (9.7)$$

Therefore:

$$L/R = d/aV \qquad (9.8)$$

and:

$$d/V = aL/R \qquad (9.9)$$

Substituting (9.8) and (9.9) into (9.4):

$$S = \frac{d/aV}{aL/R + d/aV} \qquad (9.10)$$

which simplifies to:

$$S = \frac{1}{1+a} \qquad (9.11)$$

From (9.11) it is clear that throughput has a maximum value of 1 and is inversely proportional to a. Figure 9.9 illustrates the effect of a upon throughput, S. When $a = 0$, S increases in direct proportion to G until 100% utilization is achieved. Any further increase in G results in saturation. Any increase in a, as expected from (9.11), causes saturation to occur at a lower value of S. It is clear from the graphs that $1/(1 + a)$ forms an upper bound upon utilization or efficiency, irrespective of the access control mechanism used by the network.

Since an increase in a decreases network throughput, design should attempt to make a as small as possible. One method of achieving this is to keep the medium

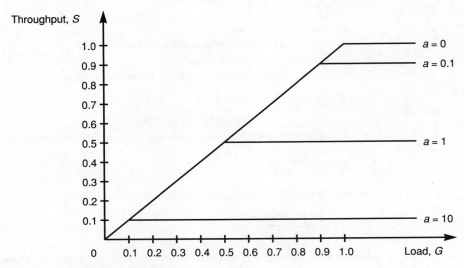

Figure 9.9   Throughput as a function of a.

length as short as possible. There are clearly practical limits to this. Alternatively, the effect of *a* can be reduced by increasing the length of a frame. However, if messages are appreciably shorter than the frame length no advantage results. A disadvantage of long frames is that transmission delay is increased and, if collisions occur, the actual load to be carried by the network may increase. Ideally, messages should be equal to an integral number of frames to optimize efficiency. Then no short frames occur and throughput is maximized.

The foregoing assumes that the maximum propagation time always occurs. This is patently not the case for every transmission for the particular type of network topologies discussed. For instance, the propagation time between two adjacent station in a token ring using RAR is considerably shorter than the maximum. Only one frame transmission at a time is also assumed. Broadband systems, which will be discussed in Section 9.3, can enable two or more simultaneous frames. However, any more favourable conditions in distance between stations and multiple frame operation tends in practice to be mitigated by an increase in the overhead imposed by the requirement of a more complex MAC protocol.

9.2.3 Protocol performance

To compare different types of LAN, a valid basis for comparison, or model, must be established. The model we shall use assumes the following:

- N active stations connected to the network.
- Propagation time is normalized such that it has a maximum, or worst-case value, equal to a.
- Each station is always prepared to transmit a frame. Although not true in practice, this does enable a maximum value of throughput S to be determined and is one of the major considerations in both design and operation of a network.

Token ring performance

Ring activity, in both token ring and token bus networks, alternates between frame transmission and token passing. Consider the case of a single data frame followed by a token which we can regard as a cycle. First, some definitions: Let T_1 be the average time taken for a station to transmit a data frame; and T_2 be the average time elapsed between one station gaining a token and the next station receiving a token. T_2 can also be regarded as the cycle time or the average time required per station to complete the transmission of a frame.

From (9.3) it is clear that normalized system throughput may be expressed as:

$$S = \frac{T_1}{T_2} \tag{9.12}$$

That is the time required to transmit one frame expressed as a fraction of the total time that the medium is occupied, or cycle time. Now consider the effect that *a* has upon throughput. For the analysis, frame duration T_1 will be normalized to 1 and therefore, from (9.1), propagation time becomes equal to *a*.

First, consider the case when the medium is physically short in relation to the duration of a frame, thus a is less than 1. In this case, the transmitter receives the start of its frame *before* it has finished transmitting the end of the frame. We shall assume that we are using RAT in which case the station releases a token immediately it ceases transmission of its frame. Hence the time elapsed between receiving a token and releasing a token is T_1 and equal to the normalized time interval of 1. The token is received by the next station after some time interval owing to propagation between stations. The average value of time to pass a token between stations is a/N. Hence $T_2 = T_1 + a/N$. Substituting into (9.12) yields:

$$S = \frac{1}{1 + a/N} \qquad a < 1 \tag{9.13}$$

Now consider a ring with medium length where propagation delay exceeds that of a frame, that is, a greater than 1. A transmitting station receives the start of its frame after the frame has propagated the full distance of the ring, that is, after time interval a, since propagation time is normalized to a. The average time to pass the token is, as before, a/N. S now becomes:

$$S = \frac{1}{a + a/N} \qquad a > 1 \tag{9.14}$$

$$= \frac{1}{a(1 + 1/N)} \qquad a > 1 \tag{9.15}$$

We may now compare (9.13) and (9.15). When the medium is physically long, or frames are short, a is greater than 1. Throughput, if N is reasonably large, is to a close approximation degraded by the factor a compared to relatively short medium lengths or long frames, when a is less than 1. It is not surprising that throughput diminishes with increasing medium length. A new frame cannot successfully be applied until the current frame has traversed the medium. The longer the medium, the lower the rate of application of successful frames, hence throughput is reduced.

CSMA/CD performance

The analysis of throughput for this MAC technique is more complex than for token ring and relies upon probability theory. Such an analysis is beyond the scope of this text but it may be shown (Stallings, 1991, pp. 414–18) that in the limit, as the number of stations N becomes very large, S has a maximum possible value:

$$S_{max} = 1 \qquad a < 1 \tag{9.16}$$

Or:

$$S_{max} = 1/a \qquad a > 1 \tag{9.17}$$

Some typical values of a are shown in Table 9.1. Note that where signalling rate is less than 1 Mbps or medium length is under 1000 m, then the value of a becomes negligible.

Figure 9.10 shows throughput for both types of MAC with various values of a (Figure 9.10a) and numbers of stations (Figure 9.10b). As might be expected

Table 9.1 Relationship between signalling rate, frame size, medium length and transmission coefficient.

Signalling rate (Mbps)	Frame size (bits)	Medium length (m)	Transmission coefficient, a
1	100	1 000	0.05
		10 000	0.5
	1 000	1 000	0.005
		10 000	0.05
	10 000	1 000	0.0005
		10 000	0.005
10	100	1 000	0.5
		10 000	5.0
	1 000	1 000	0.05
	1 000	10 000	0.5
	10 000	1 000	0.005
		10 000	0.05
50	100	1 000	2.5
		10 000	25
	1 000	1 000	0.25
		10 000	2.5
	10 000	1 000	0.025
		10 000	0.25

intuitively, token LANs outperform CSMA/CD under almost all circumstances. Ultimately the choice of LAN is based upon the trade-off between performance and complexity.

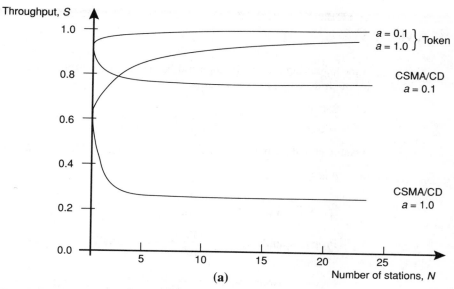

Figure 9.10 Throughput comparison of CSMA/CD and token: (a) Values of *a*.

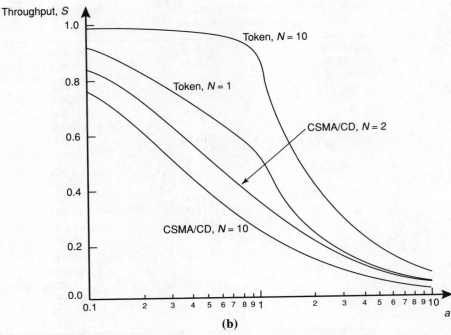

Figure 9.10 *continued* Throughput comparison of CSMA/CD and token:
(b) Numbers of stations. Source: Stallings, William, *Data and Computer Communications*, 4th edn, © 1994, pp. 416, 417. Reprinted by permission of Prentice-Hall, Upper Saddle River, New Jersey.

EXAMPLE 9.2

If the signalling rate is 10 Mbps and frame length 2000 bits, calculate the value of a for the following medium lengths:

(a) 250 m

(b) 4000 m

Assume signal propagation, typical of twisted-pair cable, to be 2×10^8 m/s.

(a) a is given by:

$$a = \frac{\text{propagation delay}}{\text{frame transmission time}}$$

$$\text{Propagation delay} = \frac{250 \, \text{m}}{2 \times 10^8 \, \text{m/s}} = 1.25 \, \mu s$$

$$\text{Frame transmission time} = \frac{\text{no. of frame bits}}{\text{signalling rate}} = \frac{2000 \, \text{bits}}{10 \, \text{Mbps}} = 200 \, \mu s$$

Therefore:

$$a = 1.25/200 = 0.00625$$

(b)   If the medium length is changed from 250 m to 4000 m then a is increased by 4000/250 or 16 times. Hence a becomes 0.1.

9.3 Broadband operation

LANs commonly apply data directly to the transmission medium as a d.c. signal. Alternatively, or as an optional variant within a standard, data may be modulated, generally by means of FSK, onto a high-frequency RF carrier and then applied to the medium. This is known as **broadband transmission**. A variety of frequencies are used, ranging from a few to several hundred megahertz. Broadband systems generally use readily available cable TV (Community Antenna Television, or CATV) technology and components. The use of RF signalling enables propagation over appreciable distances, of the order of tens of kilometres, with the aid of simple and cheap amplification. Although d.c. signalling is possible over comparable distances, the equivalent electronics required for periodic regeneration of signals is complex and prohibitively expensive for LAN applications.

There are a number of broadband modes of operation. Figure 9.11(a) shows how, since amplifiers are inherently unidirectional, two separate transmission paths may be provided. At one end, the **headend** of the network, the two buses are passively connected together to pass transmit frames to the receive path.

Since modulation is employed, there is no reason why multiple carriers, or frequency-division multiplexing, cannot be used. By using two or more separate frequencies, signals may readily be passed in both directions, Figure 9.11(b). The headend must then be active and retransmit the incoming signal at its output at a different carrier frequency by means of frequency conversion. Unlike a passive headend arrangement, a single bus may now be used but the difficulty of producing

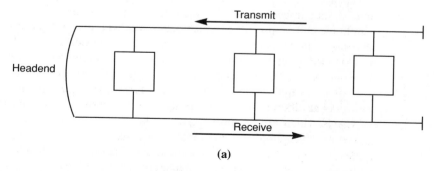

(a)

Figure 9.11   Broadband operation: (a) passive headend.

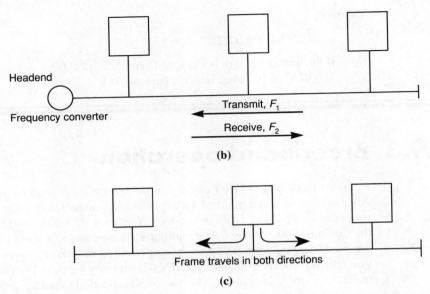

(b)

Frame travels in both directions

(c)

Figure 9.11 *continued* Broadband operation: (b) active headend; (c) carrierband.

two-way amplifiers limits the position of an amplifier to within the headend arrangement.

Carrierband operation, Figure 9.11(c), is a cut-down, cheaper version of broadband operation where only a single carrier frequency is used. Such systems are generally bidirectional over a bus topology, do not have a headend or amplifiers and hence are suitable only for shorter medium lengths.

9.4 **Repeaters, bridges and routers**

A **repeater**, in its simplest form, simply reshapes and retimes data and then retransmits it. Its purpose is to restore the integrity of the signal in regard to pulse shape, amplitude and timing, all of which deteriorate with distance. In this way the length of a baseband transmission medium, and hence physical size of the network, may be extended to facilitate a larger number of users. A repeater is unintelligent, merely repeating each bit that it receives, and can therefore be regarded as transparent.

The layered nature of the OSI reference model enables interconnection of networks at different layers. A repeater is a device used simply for interconnection of LANs which are of similar type at the physical layer. Stations connected to networks which employ a ring topology are usually active and therefore effectively perform a 'repeater' function automatically. Repeaters are therefore mainly found in bus-based networks where they interconnect numbers of bus sections, known as **segments**, to extend the effective medium length beyond the basic specification.

Segments may vary in regard to the type of channel, for example a repeater may connect a coaxial bus to a fibre-based bus. However, because repeaters are transparent, frames generated on any segment are passed, via repeaters, to every other segment. This can be inefficient in that frames do not always need to be presented to every segment. As a result the propagation time associated with frame transmission and the load on each segment is increased, which leads to some reduction in throughput, and hence loss of network performance.

Bridges are used to connect LANs , which may or may not be similar, at the data link layer. Bridges differ from repeaters in that they are intelligent devices which examine each frame that they receive to perform a 'filtering' function; MAC frame addresses are examined and only those frames containing addresses which indicate that they are intended to pass between networks are accepted by a bridge. All other intranetwork frames are rejected by the bridge. In order to perform such filtering, frames are first buffered before any possible onward transmission. Such processing can add an appreciable delay to frames passing between networks via a bridge. Frames which are permitted to pass between networks, as with a repeater, are unaltered. A bridge may be regarded as performing a frame relay function.

As a LAN expands in size, repeaters may be introduced to accommodate more users. The disadvantage is that performance deteriorates, with the number of stations because every frame is passed by repeaters to every segment, increasing overall loading of each segment. Where usage becomes high, the network may be separated into two or more identical LANs interconnected by a bridge or bridges to improve network efficiency. Splitting a LAN into several smaller LANs must be done carefully to ensure that users are grouped, as far as possible, into separate communities of interest such that their communications requirements with other LANs do not generate excessive bridge traffic. Each LAN is now required to support only its own level of traffic and the occasional appearance of traffic in the form of internetwork frames via a bridge. Should bridge traffic become excessive, frames may arrive at a bridge more quickly than it is able to respond leading to a bottleneck and an attendant degradation in service.

There are applications, other than those purely related to loading and performance, which may warrant the use of a bridge. Where only one LAN exists, a fault may affect the entire network and put every station out of service. If users are connected to one of several bridged LANs, reliability may be improved since, if one LAN fails, it affects only users on that particular network. Security is another feature which may be supported by bridges. Separate LANs are commonly configured for distinct operations within a business, for example payroll, personnel records and general application software. A bridge, by examining source addresses within frames, may reject frames depending upon access privileges ascribed to a frame's associated user. In this way access to different LANs may be controlled for reasons other than merely the suppression of frame transmission to LANs not associated with the destination address.

The last example of the use of a bridge indicates that they may offer more sophisticated functions than merely an interconnection tool. A LAN may be interconnected to more than one bridge. The corollary to this is that a frame on such a

LAN may be transmitted to the destination LAN via two or more bridges. This leads to a number of difficulties, not least of which is frame duplication. Where multiple bridges exist on a LAN, each bridge requires knowledge of the wider interconnection arrangements and a strategy is required to ensure efficient passing of frames around the network and guard against frame duplication.

As already indicated, bridges may interconnect dissimilar LANs. Two approaches are found in practice. **Encapsulating bridges** enable interconnection of similar networks via an intermediate LAN of different type. In order to perform such a bridging operation the end-user frames are encapsulated into the frames that the bridge passes to the intermediate network. Alternatively, a **translational bridge** enables direct interconnectivity between dissimilar networks. This is achieved by the bridge changing frame structures from those received by one network into suitable form for transmission to another dissimilar network.

Routers operate at the network layer and are used to interconnect similar or, more usually, dissimilar networks. Although bridges and routers assume that the upper four layer protocols are the same, a router makes use of network layer addresses rather than MAC layer addresses used by a bridge within the data link layer. Routers differ from bridges in that they typically perform LAN–LAN interconnect via a WAN, that is, LAN–WAN–LAN interconnection, whereas bridges effect the same but do not employ an intermediate WAN or have any need to involve themselves with network layer protocols.

A router enables a packet switched network to be formed by interconnecting a number of subnetworks, each of which is generally a LAN. Each subnetwork is connected to a single router and routers are then interconnected. In practice a router is necessary only where three or more subnetworks exist, and to facilitate reliability each router is ideally connected to at least two other routers, via links of a WAN, to provide alternative routing strategies. It therefore follows that a router is also required to generate the routing information for intermediate routers, when necessary.

9.5 LAN standards

Earlier sections of this chapter explored a range of MAC protocols for use with LANs and also made some performance comparison. This section explores the IEEE standards for use with LANs and MANs.

9.5.1 IEEE 802.2

The IEEE 802.3, 4 and 5 standards primarily specify details relating to the respective MAC protocol and various layer 1 arrangements, for example, details relating to the medium such as transmission rate. IEEE 802.2, as discussed in Chapter 1, is a standard common to the three LAN standards above and is a Logical Link Control (LLC) protocol responsible for addressing and data link control. Figure 9.12 shows

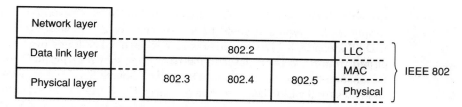

Figure 9.12 IEEE 802 and OSI.

the three layers found in the IEEE 802 standards and their association with the ISO reference model layers.

Figure 9.13 illustrates how an N-layer PDU (N-PDU) is passed from the network layer via the LLC layer to the MAC layer where a frame is formed for transmission onto the medium. The LLC layer provides a data link control function, much of which is based upon HDLC (see Chapter 7). Three types of service are offered:

(1)   *Unacknowledged connectionless mode*   No flow or error control. If required, flow and error control may be added at a higher layer.

(2)   *Connection-oriented (CO) mode*   Similar to that offered by HDLC whereby a logical connection is offered with error and flow control.

(3)   *Acknowledged connectionless mode*   A datagram service, similar to (1), but with acknowledgement of frames.

All three types of service use a standard LLC PDU format shown in Figure 9.14.

Only the unacknowledged connectionless mode will be described since this is the mode used by almost all LANs, especially those found in office and technical environments. In connectionless mode there is no necessity to establish a connection. Hence the only user service function provided by the network layer is an

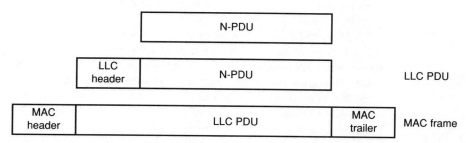

Figure 9.13   LLC layer and MAC frame.

DSAP	SSAP	N-PDU

Figure 9.14   LLC PDU.

L-DATA.request which, in OSI parlance, is known as a **primitive** (a primitive is simply a PDU type and could be a command, or a response, for example). The L-DATA.request N-PDU primitive is used to send a data frame over a network. It is passed from the network layer to the LLC layer shown earlier in Figure 9.12. The Destination Service Access Point (DSAP) and Source Service Access Point (SSAP) addresses for the LLC layer are then obtained. The N-PDU is added to form the LLC PDU which is then passed to the MAC layer where it becomes a MA.DATA.request MAC primitive. Parameters associated with the MAC primitive are:

(1) MAC layer Destination Address (DA) and Source Address (SA),

(2) service class, for example token priority,

(3) user data field length indicator.

A frame is prepared and transmitted and consists of additional fields dependent upon the IEEE 802 LAN specification in use. We shall look at the MAC frames in later sections of this chapter where we shall consider IEEE 802.3, 4 and 5 in some detail.

It should be added here that for a single LAN the network layer's main functions, that is establishment of a connection and routing, are not necessary. A station originating a MAC frame transmission includes its own address and of course that of the intended recipient. The latter address is largely derived from the user's instructions. For instance, selection of an e-mail user to whom a message is to be sent implicitly provides the required address of the receiving station. Thus we see that many LANs resolve addressing and routing in a simple manner by means of LLC and MAC peer-to-peer protocols broadly implemented at layer 2 of the ISO reference model.

9.5.2 IEEE 802.3 CSMA/CD

IEEE 802.3 which, for historical reasons, is also known as Ethernet, is arguably the most popular type of LAN currently in use. It employs a CSMA/CD MAC protocol similar to that discussed in Section 9.1.3.

The standard defines a range of options for the physical media shown in Table 9.2. The basic standard allows for a maximum segment length of 500 m and a total bus system not exceeding 2500 m. A **segment** is merely an un-repeatered section of cable. This gives rise to the **nonrooted branching tree** structure shown in Figure 9.15. It is 'nonrooted' since no headend or master exists. Segments not exceeding 500 m may be interconnected via repeaters. In this way unacceptable signal deterioration in networks up to the maximum system length of 2500 m is avoided.

With reference to Table 9.2, IEEE uses certain abbreviations for the various CSMA/CD derivatives. Originally IEEE 802.3 only specified the use of 50 Ω coaxial cable of 0.4 inch diameter at up to 10 Mbps. Baseband transmission is used and the maximum segment length of the bus is set at 500 m. Branch cables of each station are interconnected to the bus using a device known as a **tap**. Station taps are at multiples of 2.5 m with respect to each other to ensure that any signal reflection which they may introduce ensures that signal levels are a peak at all other stations.

Table 9.2   IEEE 802.3 physical medium variants.

	Transmission medium	Signalling technique	Data rate (Mbps)	Maximum segment length (m)
10Base5	Coaxial cable (50 Ω)	Baseband (Manchester)	10	500
10Base2	Coaxial cable (50 Ω)	Baseband (Manchester)	10	185
10Base-T	Unshielded twisted pair	Baseband (Manchester)	10	100
10Broad36	Coaxial cable (75 Ω)	Broadband (DPSK)	10	3600
10BaseF	Fibre	N/A	10	2000

The above specification is described as 10Base5. A lower cost system suitable for PC networks, dubbed Cheapernet, was later specified with 0.25 inch coaxial cable. Such cable is more easily installed as it is physically more flexible. This poorer quality cable is limited in use to 185 m segments and fewer taps. This system is termed 10Base2 and limits the physical area which may be served. Unshielded twisted-pair operation (UTP), utilizing standard telephone cable, is also specified

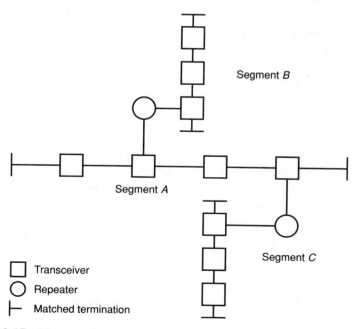

Segment *B*

Segment *A*

Segment *C*

☐ Transceiver
◯ Repeater
⊢ Matched termination

Figure 9.15   Nonrooted tree structure.

by IEEE. This offers an even cheaper medium, especially if existing cable pairs are used. An initial version known as StarLAN appeared operating at 1 Mbps reflecting the lower performance of twisted-pair compared to coaxial conductor. As its name implies, stations were connected in star fashion, a pair for each direction of transmission, to a central hub. Star topology is more suited to traditional telephone cable installations. A 10Base-T system has also appeared with a maximum segment length of 100 m using simple and cheap twisted-pair conductor and is proving very popular.

Figure 9.16 indicates the frame structure used by IEEE 802.3. The preamble, in conjunction with Manchester line coding, discussed in Chapter 4, provides a period during which the receiver may synchronize its clock with that of the incoming bit stream. Once synchronized, a receiver monitors the incoming bit stream for the Start of Frame Delimiter (SFD) pattern from which the receiver may then correctly align itself with the incoming frame structure. Destination and source addresses (DA,SA) may be defined as either 2 or 6 octets long. Full 6-octet addressing enables global, rather than local, addressing to be used.

The length field indicates to the receiver how many data bytes to expect. The data field, which carries the LLC PDU, depends upon the message size and is therefore conveniently chosen to be of variable length up to a maximum of 1500 bytes. To ensure that the frame length is at least equal to twice the maximum propagation delay for reasons discussed in Section 9.1.5, the pad and data fields must be equal to at least 46 bytes. Where there are less than 46 bytes of data to be transmitted within a frame a pad field is provided. The pad field adds sufficient bytes to ensure that the data and pad field lengths in combination equal 46 bytes.

Finally, 32 check bits are determined by dividing the data (that is both address fields, length, data and pad) by a cyclic code generator polynomial. The remainder of this division constitutes the check bits which are then placed in the frame check sequence (FCS) field. This process is known as a cyclic redundancy check or CRC. The following generator polynomial is used to perform an extremely thorough check which can detect bursts of errors of up to 31 bits:

$$x^{32} + x^{26} + x^{23} + x^{22} + x^{16} + x^{12} + x^{11} + x^{10} + x^8 + x^7 + x^5 + x^4 + x^2 + 1$$

The CSMA/CD access mechanism was described in some detail earlier in the chapter. IEEE 802.3 does not provide a separate network management function for the detection and recovery from abnormal network conditions. Rather, each station is equally responsible for network operation, hence control is fully distributed. This also means that no initialization procedure is required to add or subtract stations.

Preamble	SFD	DA	SA	Length	Data	Pad	FCS
7	1	2 or 6	2 or 6	2	0–1500		4

Number of bytes

SFD Start of frame delimiter   SA   Source address
DA   Destination address   FCS Frame check sequence

Figure 9.16 IEEE 802.3 frame.

9.5.3 IEEE 802.4 Token bus

This LAN is less common, being mostly used in automated factory applications. Physical layer specifications include a number of coaxial-based media derived from CATV technology with the idea of utilizing readily available, low-cost components. Table 9.3 shows some token bus variants.

Although a bus topology is used, a logical ring is established. Stations are arranged in a ring by declaring, for a given station, the address of the next station as a **successor** and that of the preceding station a **predecessor**. Tokens and frames then circulate around the ring passing from each station to its successor in turn.

Once a ring is established, further stations may be added to the ring. Any station holding the token may send a solicit_successor frame at random intervals in time. This frame specifies a range of currently inactve addresses between that of the sender and its successor. Any station whose address is within this range may respond within a predetermined interval equal to the slot time of the ring, or **response window**. Because the system is actually a bus topology, more than one station waiting to join the ring may respond. If this occurs the frame transmissions on the medium become corrupted. A contention arbitration procedure is then invoked, using a resolve_contention frame, the result of which is that only one station initially succeeds in becoming a successor. All other stations in contention then wait for the next round of soliciting a successor.

Initialization of the ring is a special case of adding stations. Any station that detects that the ring is idle for too long transmits a claim_token. As with the solicit_successor frame, two or more stations may issue a claim_token at once. A mechanism also exists here to resolve such contention which results in only one station becoming responsible for initialization of the ring. This station is at this moment the only operational station connected to the ring. It now embarks upon a process of enabling another station to join the ring by issuing a solicit_successor frame. The ring then builds itself by connecting further stations in the manner already described. This action, although used when a network is first powered up, may also be used for fault management purposes, for example if part of the bus has been disconnected or a station has been removed.

Table 9.3 IEEE 802.4 token bus physical medium variants.

	Transmission medium	Signalling technique	Signalling rate (Mbps)	Maximum segment length (m)
Broadband	Coaxial cable (75 Ω)	Broadband (AM/PSK)	1, 5, 10	Not specified
Carrierband	Coaxial cable (75 Ω)	Broadband (FSK)	1, 5, 10	7600
Optical fibre	Optical fibre	ASK-Manchester	5, 10, 20	Not specified

Having established the principle of creating and maintaining the ring, we shall now consider access control under normal operation. Two phases of operation exist, namely token transfer and data transfer. A token is released after the last station has completed its transmission (RAT). The token then passes to its successor which may, if it desires, transmit a message via one or more data frames until it has either completed its message, or timed out. If the successor has no frames to send it retransmits the token.

After a station passes the token, it listens to ensure that its successor sends a valid frame so indicating that the token was successfully received. Otherwise remedial action then follows which, in the extreme, can result in the ring having to be re-established. Where a successor fails to make a transmission within a certain time it may be assumed to be inactive or defective. The station's predecessor then sends a who_follows_me frame with its successor's address in the data field. Each station receiving this frame compares the address contained therein with that of its own predecessor's address. The station for whom the address is that of its predecessor responds by sending its own address in a set_successor frame and hence becomes the new successor.

If no new successor is found, the original station attempting to pass a token sends a solicit_successor frame containing its own address. Any station may respond and become a successor. If a successor is not produced by this procedure, then the system has virtually collapsed and the original station that sent a token monitors bus activity until some future time when it appears that the ring may be rebuilt. This is indicated by suitable frames, for example claim_token from another station, which would indicate an attempt to rebuild the ring.

The frame structure for token bus is shown in Figure 9.17. The preamble is followed by a Start of frame Delimiter (SD). Although similar to that of IEEE 802.3 it differs in that rather than specifying length, an End of frame Delimiter (ED) is used. Additionally, there is a Frame Control (FC) field. Two FC field bits indicate:

00   MAC control frame
01   LLC data frame
10   Station management data frame
11   Special-purpose data frame

In general, frames are either MAC or LLC data frames and therefore station management and special-purpose frames will not be discussed. Where a MAC frame is transmitted then six further bits are used to indicate:

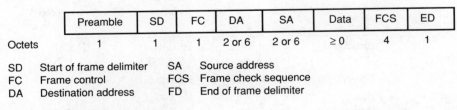

Preamble	SD	FC	DA	SA	Data	FCS	ED
Octets	1	1	1	2 or 6	2 or 6	≥ 0	4	1

SD	Start of frame delimiter	SA	Source address
FC	Frame control	FCS	Frame check sequence
DA	Destination address	FD	End of frame delimiter

Figure 9.17   Token bus frame.

claim_token
solicit_successor
who_follows
resolve_contention
token
set_successor

When the frame is LLC data, three bits enable other stations to be polled and responses returned. This is achieved by transmitting a request_with_response data frame. When a station receives such a frame it may respond with a response data frame after which the right to transmit returns to the original station. Alternatively, a request_with_no_response may be sent.

Token frames may specify one of four levels of priority using three bits of the FC field. When a station receives a token it sends any waiting high priority frames until a **High Priority Token Hold Timer** (HPTHT) associated with the station expires. If, after all the high priority frames have been sent, any lower priority frames are awaiting transmission these may be sent provided adequate time exists. Stations have another timer called the **Token Holding Timer** (THT) whose value is computed as follows. Each station has a **Token Rotation Timer** (TRT) which times the interval that has elapsed between a station receiving a token and the last time it did so. Upon receiving a token, the value of TRT is placed into THT and TRT reset and, as mentioned above, any high priority frames are transmitted. The difference between the value of THT upon receipt of a token and a fixed time, known as the **Target Token Rotation Time** (TTRT), is computed and provided it is positive, any lower priority frames may be transmitted once high priority frame transmission has ceased. Low priority frames may be transmitted for a period of time not exceeding the computed difference between THT and TTRT. Both priority and fair share operation may be achieved in the above manner.

Fault management has already been alluded to. There are also other fault conditions to consider, for example duplicated tokens. Note that token bus, as with IEEE 802.3, is a distributed control system. Each station has equal power and responsibility for the correct operation of the ring, there being no common monitor function.

9.5.4 IEEE 802.5 Token ring

Token ring is a very popular LAN and one of the oldest, first proposed by IBM in 1969. It is commonly found in commercial environments. The IEEE 802.5 standard enables up to 250 stations to be connected with multiple levels of priority.

Token ring uses shielded twin twisted-pair cable with baseband transmission at either 1, 4 or, more recently, 16 Mbps. Stations are connected to the medium via a trunk coupling unit which enables inactive or failed stations to be bypassed. Symbols are transmitted using differential Manchester encoding to enable receivers to synchronize with the incoming bit stream.

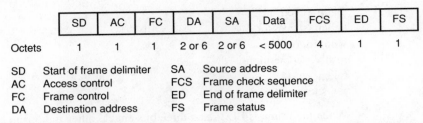

SD	AC	FC	DA	SA	Data	FCS	ED	FS

Octets   1   1   1   2 or 6 2 or 6 < 5000 4   1   1

SD   Start of frame delimiter   SA   Source address
AC   Access control   FCS Frame check sequence
FC   Frame control   ED   End of frame delimiter
DA   Destination address   FS   Frame status

Figure 9.18 Token ring frame.

When all stations are idle, a short token circulates around the ring and the token is labelled as **free**. A station wishing to transmit awaits the arrival of a token. The station then inserts its own data into the frame, marks it as **busy** and immediately transmits. Busy frames are removed at the transmitter (RAT) and a free token is then re-applied to the medium. In the absence of a free token, no other station is able to access the ring.

The token ring frame is shown in Figure 9.18. Note that, unlike Ethernet and token bus, no synchronizing preamble is necessary because there is a constant stream of bits circulating the ring.

Two types of frame may be transmitted, namely token or normal.

Token frame

This consists of a subset of the full token ring frame and is shown in Figure 9.19. A token frame contains SD, AC and ED fields. The ED field contains an (Intermediate) I-bit which, if set to 1, indicates that it is an intermediate frame and that another frame from the same transmitter will follow. A 0 indicates that the frame is either the only frame, or the last. The AC field, which has 8 bits, is of particular interest and is shown in Figure 9.20. If the token bit is a 0, then it is a token. Otherwise a 1 indicates a data (or normal) frame.

Two sets of three bits, known as Reservation (R) bits and Priority (P) bits, permit eight levels of priority, 0 through 7, the latter being the highest. A station with a particular priority (priority being determined by a higher layer protocol) wishes to send a frame and waits for a token. Upon receipt of a token, the priority bits are examined. Providing the token's priority is not higher than that of the station, the station may seize the token. The station transmits an information frame (I-frame) with the reservation bits set to 0.

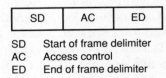

SD	AC	ED

SD   Start of frame delimiter
AC   Access control
ED   End of frame delimiter

Figure 9.19 Token frame.

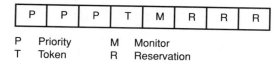

P   Priority   M   Monitor
T   Token   R   Reservation

Figure 9.20   Access control field.

A station which is successful in gaining the token may continue to transmit frames until either it has no more frames to send, or its THT expires (the default holding time is 10 ms), whichever occurs first. A free token is then released by the station with:

(1)   priority bits set equal to the reservation bits, and

(2)   reservation bits replaced by the priority bits of the station, if it has a higher priority.

If a waiting station receives a free token which has a higher priority than itself, the station then examines the reservation bits of the token and, if the station has a higher priority, it places its own priority into the reservation bits. The token is then passed on.

It is apparent that token priority, indicated by priority bits, may so far only be increased, not decreased. In order to overcome this problem, and so allow lower priority stations to transmit eventually, an additional rule must be strictly adhered to. A station raising token priority must eventually restore it to its original value. This is achieved by a station which raised priority storing the priority bits on a **stack**. Such a station is known as a stacking station. A stacking station will eventually receive a token of the same priority as the value to which it raised an earlier token. It may then be assumed that there are no further stations awaiting transmission at this, or higher, priority. Then, if the stacking station has no frames to transmit, it unstacks the previously stored, lower priority and places it in the priority bits. There may be a number of stacking stations active within a ring at any one time where priority bits are successively raised by a number of stations in turn.

To summarize this priority strategy: a station may only seize a token of given priority if it has the same or higher priority. While a station is actively transmitting, other stations may bid for the next free token to be released by indicating their priority using reservation bits. Tokens are released with priority set to that of the waiting station with the highest priority in the ring. Only the highest priority station is able to seize a token. Over time high-priority stations complete their transmissions and the priority bits become replaced by lower values, obtained from the reservation bits, enabling lower priority stations to transmit.

Normal frame

Normal frames, containing all fields, are used to send either data or MAC information around the ring once a station gains a token.

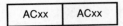

Figure 9.21 Frame status field.

An FCS field provides a 32-bit CRC error check. The FC field indicates that it is either a MAC or I-frame. MAC information relates to the following:

- usage of the token
- control of priority messages
- management of network in the case of errors or failure

All stations interpret and, if necessary, act upon MAC information. In the case of an I-frame, control bits within the FC field are interpreted only by those terminals identified by the DA field. Although the token access control is one of the MAC functions, there are a number of other functions necessary for correct operation of the ring which include initialization of the ring and, designating another station as a backup to the active monitor (to be discussed shortly).

The Frame Status (FS) field, consisting of two 4-bit nibbles, is shown in Figure 9.21. The originating station sets both A (addressed-recognized bits) and C (frame-copied bits) to zero. Any station which has the address, or is within the range of addresses, specified by a MAC I-frame sets the A bits. Station/s for whom an I-frame is addressed may also copy the frame for further processing and set the C bits. Hence when a frame returns to the originating station it is possible, by examination of A and C, to determine if the addressed station/s are:

- inactive
- active but did not copy frame
- active and copied frame

Ring management is achieved in the following manner. Each station is active and operates as a repeater. Error detection and recovery facilities are performed by one of the stations functioning as a monitor. Any station has the potential to become a monitor. This also enables backup in the event of a monitor failing.

The active monitor periodically signals its presence by transmitting an Active Monitor Present (AMP) MAC frame. The first station to receive the AMP becomes the standby monitor and resets a timer and sends a Standby Monitor Present (SMP) frame. If at any time the standby's timer expires, it is taken as an indication that the active monitor has failed. The standby becomes the active monitor and then sends an AMP to produce another standby monitor.

The monitor also provides a latency buffer. Since the token consists of 24 bits, the ring must be at least 24 bits long. In physically short rings, this may not be the case. The latency buffer introduces additional bits as necessary to pad out the ring. The monitor also provides the master oscillator for the network from which all other stations are bit synchronized.

The active monitor also checks the ring for correct transmission. There are two main error conditions:

(1)   no token or frame circulating

(2)   persistent busy frame

Absence of a token or frame is detected by the monitor which times the interval between tokens and frames. This interval must not exceed the propagation time of the ring, if it does a fault has occurred. The monitor then purges the ring of all frames and initiates the generation of a new token.

Under normal operation, each busy token frame passing through the monitor has a **monitor bit** set to 1. It is reset by the originating transmitter. A continuously circulating busy frame would arrive at the monitor with the monitor bit already at 1. In such cases, the monitor changes the busy token to free.

Other functions of the monitor are to enable the establishment of the ring and enable stations to be added or removed from the ring. If the ring has collapsed, for example if the medium is severed, no frames circulate. A station, upon discovering this condition, sends a beacon frame. When the ring is restored, a beacon frame returns and the ring may then be reinitialized.

9.6 Fibre distributed data interface

FDDI was developed by ANSI as a high-speed LAN operating at 100 Mbps. It has been defined in ISO 9314 and uses optical fibre. It is a popular **backbone**, used to interconnect a variety of lower speed LANs. FDDI may, with its relatively high transmission speed, also be used as a MAN to connect a number of urban sites.

FDDI employs two counter-rotating rings for reliability, with each using a token-based protocol. The dual ring arrangement consists of a primary and secondary ring. The secondary ring may be used in the same way as that of the primary. Alternatively, it may be used purely as a backup in the event of the primary ring failing.

Stations connected to both rings are known as **Dual Attached Stations** (DASs), and are typically devices such as hubs, concentrators or servers. **Single Attached Stations** (SASs) are connected only to the primary ring and may be workstations or printers for example. Because a single SAS device does not normally require such a high-capacity medium as afforded by FDDI, they are commonly connected instead to a multiplexer which is itself a DAS. A typical FDDI network configuration may be as shown in Figure 9.22.

At the physical layer 4B5B line coding is used and this is followed by a NRZI encoder. NRZI is an extension of NRZ, based upon AMI, both of which were discussed in Chapter 4. NRZ encoding suffers the disadvantage that clock timing information in the received signal is lost if long strings of binary 1s or binary 0s are transmitted. NRZI partially overcomes this disadvantage by arranging that no signal level change occurs when transmitting binary 1s (as with NRZ) but transmission of successive binary 0s causes a transition in the signal level (differential encoding), or vice versa. Hence clock timing is lost only when transmitting long strings of binary 1s. The combination of 4B5B and NRZI coding used in FDDI

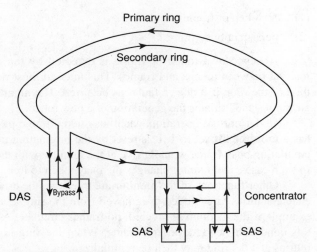

Figure 9.22 FDDI attachments.

ensures that the transmitted signal contains a transition at least once in every three bits so ensuring adequate clock timing content. Ring length can be of the order of 100 km, with up to 500 DASs or 1000 SASs. Node spacing may be of the order of a few to tens of kilometres, if repeaters are introduced.

The frame structures for a token and an I-frame are shown in Figure 9.23 and are similar to the IEEE 802.5 token ring format.

FDDI uses a protocol similar to token bus described in Section 9.5.3. There are, however, some differences. An FDDI ring may be as much as ten times the length of a token bus LAN, leading to much larger propagation delays. If FDDI

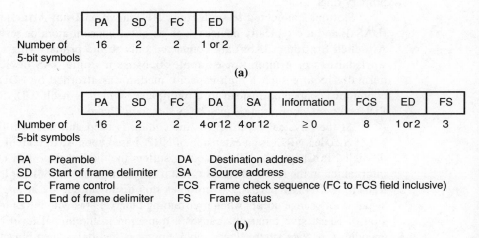

PA	SD	FC	ED
16	2	2	1 or 2

Number of 5-bit symbols

(a)

PA	SD	FC	DA	SA	Information	FCS	ED	FS
16	2	2	4 or 12	4 or 12	≥ 0	8	1 or 2	3

Number of 5-bit symbols

PA	Preamble		DA	Destination address
SD	Start of frame delimiter		SA	Source address
FC	Frame control		FCS	Frame check sequence (FC to FCS field inclusive)
ED	End of frame delimiter		FS	Frame status

(b)

Figure 9.23 FDDI frame structures: (a) token; (b) information frame.

were to use RAT, a sending station may have to wait an appreciable period of time before its transmitted frame returns to itself. Thus the time elapsed between gaining a token and its subsequent release may be very long. To improve utilization, stations transmit a token immediately after sending the FS field of an I-frame. Hence, FDDI uses what is termed an 'early release' token release mechanism. Although stations read frames addressed to them, it is the responsibility of stations initiating I-frames to remove them from the ring.

The use of early release, coupled with the relatively large size of FDDI rings, means that more than one frame may be circulating around the ring at a time. Each station relays the SD, FC and DA fields of all received frames before it may be aware that it is in fact its own frame. If the frame is a station's own frame, subsequent fields are discarded. This gives rise to the transmission of **frame fragments** around the ring. Such fragments are removed from the ring as follows. Once a station gains a token, all subsequent frames received prior to releasing the next token are assumed to be its own frames. Such frames are discarded. In this way, should any frame fragments occur after receipt of a token, they are automatically discarded as well.

Priority is also similar to that used by token bus and makes use of a TRT, a TTRT and a THT. However, FDDI differs in that it supports not only asynchronous data but also synchronous data, such as regularly occurring speech samples. Frames containing synchronous data are allocated highest priority. Each station is then allocated a synchronous allocation time (SAT) which defines the maximum length of time that a station may transmit synchronous data upon gaining a token.

EXAMPLE 9.3

An FDDI network has a maximum of 1000 single attached stations. Each station introduces 10 bits because of a buffer contained in the interface. Given that a 4B5B code is used, determine the minimum number of information bits which may be present in the ring at once.

Number of bits introduced by stations $= 1000 \times 10 = 10\,000$ bits. However, for each five bits transmitted as a result of coding, only four convey any information. Therefore there is a minimum of 8000 active information bits circulating the ring at any one time.

In practice additional bits are circulating within the transmission medium as a result of the propagation delay introduced.

Figure 9.24 shows how, if a fault occurs in a link between stations or at a station, primary and secondary rings may be reconfigured to allow continued operation.

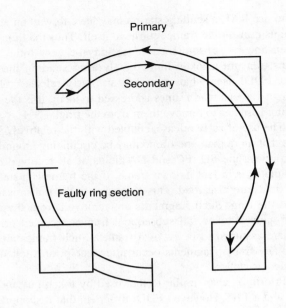

Figure 9.24 Use of dual counter rotating ring under fault condition.

9.7 IEEE 802.6 Metropolitan area network

MANs are networks comprising a number of LANs. Their geographical sizes are of the order of tens of kilometres in diameter. Interconnection of LAN sites in the highly regulated environment of WANs has traditionally been possible only via leased lines provided by PTTs. This is now changing as other operators, such as cable companies, are entering the marketplace. FDDI (see Section 9.6) was originally conceived as a LAN. It may nevertheless perform a MAN facility, but the cable infrastructure required is not standard to that of the PTT domain. A MAN using FDDI technology could be built by considerable tailoring on the part of a PTT or by the installation of a purpose-built cable network by some other service provider. Such approaches to the development of a MAN are expensive, cumbersome and also result in networks which are inflexible in regard to the addition and substraction of physical nodes.

A MAN standard, IEEE 802.6, appeared in 1990 specifically to address the above issues. The specification is so arranged that a MAN can be built around standard technology offered by PTTs and be capable of adapting as technology evolves. In Europe, MANs operate at speeds of 34 or 140 Mbps making use of optical links provided by PTTs. The large physical size of a MAN precludes the contention and token MAC protocols used by LANs. This is because time wasted by collision activity is, as has already been seen, a function of the length of the medium. In token ring networks, the average time for a station to gain a token is at best half the total ring propagation time. Physically large networks such as MANs result in a reduction in utilization if contention is used. Similarly, the use of a token may result in large access times or long delays, and the increased time in token propagation represents reduced utilization (note that FDDI addresses the latter point by using early release).

9.7.1 Distributed queue dual bus (DQDB)

The MAC specified by IEEE 802.6 is known as the **Distributed Queue Dual Bus** (DQDB) and is illustrated in Figure 9.25. As its name implies, there are two (counter-directional) buses to which every station is connected. Each station may send to or receive from either bus.

A frame generator is situated at the end of each bus. Frames consist of a header and a number of slots and are generated at a rate of 8 kHz. The number of slots contained in each frame is dependent on the transmission speed of the bus.

DQDB has been adapted from a protocol developed by a subsidiary of Telecom Australia. The protocol is designed to support asynchronous data such as computer data as well as the synchronous data of 64 kbps voice/ISDN channels. The latter is supported by permanently assigning a particular slot, or slots. Since the frames appear regularly at 8 kHz a connection-oriented service may be presented. Asynchronous data frames are arranged in blocks, known as **segments**, each of 53 bytes. Each segment has a 5-byte header and 48-byte information field. Once assembled, segments are then applied, via the MAC protocol, to nonassigned slots within bus frames. Note that asynchronous data segments, with their 53-byte structure, will support broadband ISDN and ATM which will be discussed in Chapter 11. This means that IEEE 802.6 MANs are well placed to form part of the access network to link LANs into the next generation of broadband WANs.

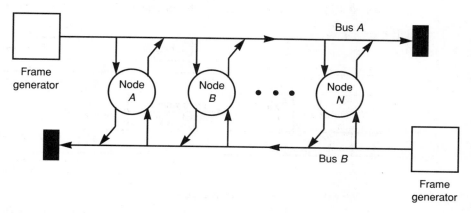

Figure 9.25   DQDB topology.

EXAMPLE 9.4

A DQDB bus operates at 34.368 Mbps. Assume that the size of the frame header is negligible in comparison to the frame's length. Estimate the number of slots contained in one frame.

Frames are transmitted at 8 kHz and therefore each frame has a duration of 125 μs. At a transmission rate of 34.368 Mbps, the duration of one bit is found:

$$\text{Duration of 1 bit} = \frac{1}{34.368 \times 10^6} = 29.1 \, \text{ns}$$

and, since each slot contains 53 bytes, we may now determine the number of slots:

$$\text{Number of slots} = \frac{125 \, \mu s}{\text{duration of one slot}} = \frac{125 \, \mu s}{53 \times 8 \times 29.1 \times 10^{-9}} = 10.13$$

Clearly the number of slots must be an integer. Hence the bus may support 10 slots per frame if the header is ignored.

Now let us examine the MAC protocol in more detail. Each node has two queues, one for each bus, into which are placed data segments waiting for transmission. Each queue has associated with it a **countdown counter** and a **request counter**. Figure 9.26 shows the two counters associated with the queue for sending data segments on bus B for a particular node. (There is also a queue and a pair of counters associated with bus A which operate in the same manner, not shown.) A node wishing to send on a particular bus makes a request on one bus and ultimately is granted a free slot on the other bus.

Nodes wishing to send a data segment on bus B set the Request bit (R) of the next slot to pass on bus A. Each R bit of a slot which is set increments the request

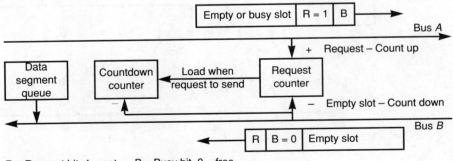

R Request bit, 1 = set   B   Busy bit, 0 = free

Figure 9.26 DQDB MAC protocol.

counter of each node as it passes. Hence the request counter's value indicates how many downstream nodes on bus *B* are queuing a data segment. Each empty slot which passes on bus *B* is subsequently used by a downstream node queuing a data segment. Such queuing nodes are in fact the same nodes that made requests on bus *A*. Therefore, each time an empty slot appears, the request counter must be decremented.

To use a free slot, a node first sets an R bit, as already mentioned, and then loads the value of its request counter into the countdown counter. The latter's value represents how many nodes are currently queuing. Each time an empty slot passes on bus *B*, the countdown counter is decremented (in addition to the request counter for reasons already stated). Once the countdown counter reaches zero, all previously queuing downstream nodes have sent their data segments and emptied their queues. When the next free slot appears on bus *B*, a data segment is then 'dequeued' and placed into it and the slot marked 'busy'. The MAC protocol operates a fair-share policy by arranging that downstream data segments awaiting transmission are sent before those of any new upstream requests are serviced.

The arrangement of a segment queue, request counter and countdown counter at each node is duplicated for both buses. Hence 'distributed queue dual bus' (DQDB). In addition, each node may have up to four levels of segment priority, each level having its own segment queue. Thus there may be as many as eight queues in total if priority is used. Synchronous data is always assigned the highest level of priority. The MAC protocol is relatively simple to implement and yet offers efficient use of the medium (using a slotted technique) as well as a fair-share access arrangement.

As with FDDI, the use of a pair of counter-directional buses enhances reliability. All nodes are able to generate frames. If the network topology is arranged as a looped bus (rather than as an open bus shown in Figure 9.25) a break in the bus can be 'healed'. Figure 9.27 shows how, simply by changing the selection of the two terminating stations, a faulty section of bus, or indeed node, may be eliminated.

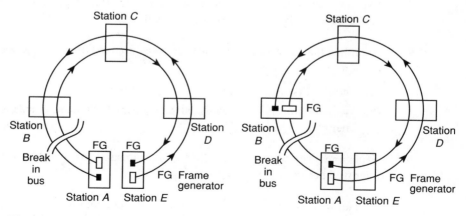

Figure 9.27　Self-healing bus.

9.8 Conclusion

This chapter commenced with Aloha, a very simple form of LAN which used the radio technology available at the time of its development. We saw how this greatly influenced further LAN developments and how some of its concepts are now established in current LAN standards. Todays LANs predominantly use coaxial cable to support the high transmission speeds required. This is now changing in a drive to push costs down. Modern electronic techniques have enabled low-cost screened twisted-pair conductors to be used instead of coaxial cable. However, at the other extreme, the expansion of LAN concepts into physically larger metropolitan areas, and their attendant very high transmission speeds, has led to the separate development of MANs. To support such networks optical fibre transmission media are necessary. Although FDDI uses only an adapted form of token ring MAC protocol, MANs have led to the development of much more sophisticated MAC protocols compared with those employed by LANs.

The wheel is now about to turn full circle with the appearance of the IEEE 802.11 standard for Wireless LANs. To overcome the installation and attendant reconfiguration problems, particularly in office-based LANs, metallic and optical media are being replaced by radio (and infra-red) transmission. This is an exciting new development which may revolutionize LAN installations in the next century. Coupled with the portability of PCs, the possibility is now emerging of users moving around within a site, complete with computer, and yet remaining actively configured to a LAN.

Exercises

9.1 Distinguish between vulnerable time and collision window.

9.2 Explain what effect frames of variable, rather than fixed length, might have in an Aloha-based LAN upon vulnerable time and collision window.

9.3 Assuming that all stations on a LAN are synchronized (for example, slotted Aloha), explain:

    (a)   how collisions may still occur,

    (b)   why the collision window is halved compared with that for an unsynchronized network.

9.4 Explain two key advantages of CSMA/CD compared with CSMA which result in improved performance.

9.5 Explain why a CSMA/CD bus may be idle, even though some stations have frames to transmit.

9.6 Explain how P-persistence algorithms used with CSMA/CD networks attempt to improve bus efficiency by comparing them with 1-persistence and nonpersistence algorithms.

9.7 Draw flowcharts to illustrate the CSMA/CD protocol, including an appropriate persistence algorithm, for:

(a) the transmitting station

(b) the receiving station

9.8 A CSMA/CD-based LAN operates at 10 Mbps with a utilization of 30%. Calculate:

(a) the number of information bits transmitted per second,

(b) the ratio between bus activity and idle time,

(c) the duration in time of a frame consisting of

(i) 18 bytes (IEEE 802.3 minimum frame length) and

(ii) 1526 bytes.

Compare each frame's length in time with the equivalent length of cable which would produce a similar time duration.

9.9 A bus-type LAN operates at 10 Mbps baseband. If the length of the bus is 450 m, calculate the vulnerable time and collision window. (Assume that signals propagate in a cable at approximately two-thirds of the speed of free-space propagation.)

9.10 An upper bound for utilization U of a CSMA/CD-based LAN may be expressed by:

$$U = \frac{1}{1+a}$$

where a is given by:

$$a = \frac{\text{propagation time}}{\text{frame transmission time}}$$

Estimate the utilization of a network with maximum length of 500 m if a frame is transmitted with:

(a) maximum size of 1526 bytes

(b) minimum size of 18 bytes

Assume a signalling rate of 10 Mbps and velocity of propagation of 2×10^8 m/s.

(c) Explain why utilization is substantially reduced with small frame sizes or long lengths of cable.

9.11 The maximum time for collision detection in a broadband bus-based system is four times the end-to-end propagation delay. Prove that this is the case with the aid of a diagram.

9.12 (a) Discuss the differences between repeaters, bridges and routers.

(b) Two separate LAN segments are to be interconnected, using either a repeater or a bridge. Although a bridge is more complex, and hence costly, what advantage does it offer over a repeater?

9.13 Explain why the CRC check field is not applied to the preamble of the IEEE 802.3 frame.

9.14 IEEE 802.3 frames contain a Pad and length field, while IEEE 802.4 and 5 token-based frames contain an ED (End-of-frame Delimiter) field only. Discuss.

9.15 Consider a slotted ring of length 10 km with a data rate of 10 Mbps and 500 repeaters, each of which introduces a 1-bit delay. Each slot contains room for one source address byte, one destination address byte, two data bytes and five control bits totalling 37 bits. How many slots may the ring support ?

9.16 Compare the relative merits of LANs using ring and bus media.

9.17 A small business uses a CSMA/CD type of LAN but finds that delays are, on occasion, excessive. To overcome this, the LAN is replaced by a token-based network. Although the usage and loading do not change, delays become less frequent and of shorter duration. Explain this improvement.

9.18 Consider the transfer of a file containing a million characters from one station to another. What is the total elapsed time and effective throughput for the following cases:

(a) A circuit-switched, star topology local network. Call set-up time is negligible and the data rate is 64 kbps.

(b) A bus topology local area network with 2 stations D apart, a data rate of B bps, and a frame size P with 80 bits of overhead. Each frame is acknowledged with an 88-bit frame before the next is sent. The propagation speed on the bus is 200 m/µs:

	D (km)	B (Mbps)	P (bits)
(i)	1	1	256
(ii)	1	10	256
(iii)	10	1	256
(iv)	1	50	10 000

(c) A ring topology with a total circular length of $2D$, with two stations a distance D apart. Acknowledgement is achieved by allowing a frame to circulate past the destination station, back to the source station. There are N repeaters on the ring, each of which introduces a delay of 1 bit time. Repeat the calculation for each case as in (b) for $N = 10$, 100 and 1000.

9.19   In token ring systems, suppose that the destination station removes the data frame and immediately sends a short acknowledgement frame to the sender, rather than letting the original frame return to the sender. How does this affect performance?

9.20   IEEE 802.4 and 5 standards both use token-based MAC protocols. Explain why the IEEE 802.4 token bus LAN has a variety of additional frames, such as 'solicit_successor', which are not found in IEEE 802.5.

9.21   In considering the frame structure of IEEE 802 standards:

(a)   Indicate those fields which are common to 802.3, 802.4 and 802.5 and briefly explain their use.

(b)   Indicate any fields which vary within the three standards mentioned, and explain their use and why such variation exists.

9.22   Explain how frame fragments may appear in FDDI networks.

9.23   An FDDI network consists of 300 single attached stations. Each station introduces 10 bits because of a buffer contained in its interface. Estimate the minimum number of information bits circulating the ring.

9.24   A DQDB bus operates at 140 Mbps. Assume that the size of the frame header is negligible in comparison to the frame's length. Estimate the number of slots contained in one frame.

9.25   Frames transmitted on a DQDB bus contain 20 slots. Calculate the transmission rate of the bus.

9.26   Discuss the features of the DQDB MAC protocol which give rise to short delays and high medium utilization.

Integrated services digital network

10

Digitalization of PSTN customer equipment and their exchange connections has made it possible to integrate a variety of nontelephony-based services with those of the more traditional telephone services. Such integrated services may now be carried over an enhanced PSTN within PTO environments known as an **Integrated Services Digital Network** (ISDN). The chapter looks at the development of the PSTN to that of an all-digital network capable of supporting ISDN operation. The supporting technology is then examined, followed by a discussion of the enhanced types of services and applications ISDN is able to offer. We shall then look at how ISDN operates for the different types of connections available. The chapter concludes with an indication of ISDN availability, both nationally and internationally.

10.1 Development of an integrated services digital network

The two major network providers in the United Kingdom are Mercury and British Telecom (BT). Mercury has, from the outset, built an all-digital network as far as switching and interexchange connections are concerned. BT is rapidly digitizing its network. Currently, most telephone exchanges and all of the trunk and junction network is digital. However, some work remains to complete the conversion from analogue to all-digital technology.

Figure 10.1 shows BT's switching plan. There are about 6000 Local Exchanges (LEs), the vast majority of which are digital. Within a small geographical locality, a number of local exchanges are interconnected by local junctions. Calls between them are connected and charged at local rate. Calls further afield are generally regarded as trunk calls and are connected via one or two **Digital Main Switching Units** (DMSUs). This switching plan is an example of a topology which contains a mixture of star and mesh configurations. There are 53 fully-interconnected or meshed DMSUs, each serving a number of LEs connected in star fashion. Dialling of the international prefix 00 automatically routes a call to the parent DMSU and on to its associated international gateway exchange although, as indicated, an alternative route to an international exchange is possible via an intermediate DMSU. In

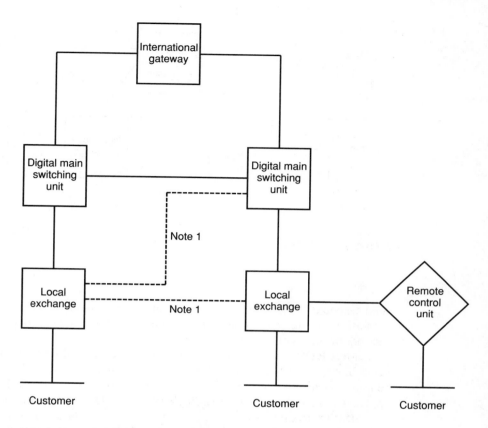

Note 1: Alternative routes where traffic usage is economic

Figure 10.1   BT switching plan.

general, the provision of alternative routes within a network provides resilience in the event of failure of either a route or a switching centre.

An **Integrated Digital Network** (IDN) is one in which all of the exchanges and their interconnections are digital. The term commonly used to describe the local line connection between exchange and customer is 'local end'. An IDN may accept analogue signals from customers over the local ends in which case signals are immediately digitized before processing at the exchange. Completion of BT's IDN will not occur before the turn of the century. Mercury's telephone service already is, and has been since its inception, an IDN.

ISDN is achieved by extending the IDN to provide end-to-end *digital* connectivity between customers which may then offer a wide range of services. Such digitization of the local end is referred to as **Integrated Digital Access** (IDA). Figure 10.2 illustrates the relationship between IDN, IDA and ISDN. Once signals are digitized, then in principle a variety of additional services may be offered. Such

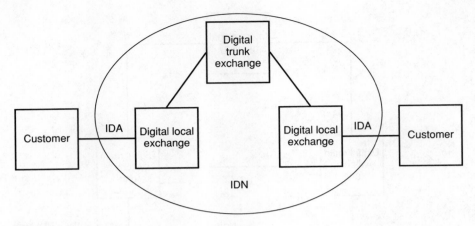

Figure 10.2 ISDN.

services could be accessed either on a dial-up basis, point-to-point or by further networking to other systems, for example by gaining access to a packet data network by means of a gateway.

Most PABXs are now digital and many are connected to the local exchange using highspeed digital links. PABXs are well placed to offer ISDN services in addition to telephony services. Such enhanced PABXs able to offer ISDN facilities are known as **Integrated Services PBXs** (ISPBXs). One such facility is the ability of an ISPBX to offer a LAN service. ISDN may also be used as a convenient switchable method of interconnecting LANs. Until very recently this has not been possible using traditional telephone networks because of the high data rates required. Rather, dedicated private systems were formed at much greater cost.

10.2 **ISDN technology**

From the outset, ISDN has sought to offer a unified approach to customers to enable simple connection of equipment to the network. This desire is borne out of early experiences of the computer industry in attempting to use the PSTN for transmission of data.

ISDN was conceived to support a wide range of services, both voice and nonvoice, to which users have access by means of a limited set of standard multipurpose customer interfaces. To support this aim a number of 'building blocks' exist:

- transparent 64 kbps channels via a universal IDN;
- control signalling external to the 64 kbps message channels (known as common channel signalling),

- standard protocols and interfaces which are, as far as is possible, independent of technology and common to many services,
- a variety of terminals which interwork with the standard network.

10.2.1 Customer access

ISDN offers a basic access to customers over a single telephone line. Basic access offers 2 B channels, each of 64 kbps, and a D channel at 16 kbps. This is known as a 2B+D service. Each B channel is full duplex and can be used to carry a single digitized voice channel or transmit data. B channels may also be multiplexed, for example to carry two 32 kbps digitally compressed voice signals. The D channel may carry data at 16 kbps. However, it is not exclusively available for data because it also acts as the out-of-band common signalling channel for normal telephony operation on B channels to pass call set-up and clear signals, and so on, between customer and exchange. However, once a connection is established, little further signalling occurs during a call leaving the D channel available for customer use.

The total bit rate of basic access is 144 kbps in each direction. Basic access supports end-to-end operation using a standard single-pair telephone line between exchange and customer. Two alternative telephone line transmission techniques exist, namely burst mode and echo cancellation.

Burst-mode operation and its frame structure are illustrated in Figure 10.3. The technique, also known as 'ping-pong', consist of forming a number of bits into bursts or frames. The D channel transmission rate is only a quarter that of a B channel, hence each frame carries four times as many B channel bits as the D channel in each burst. Frames are generated at a rate of 8 kHz and sent alternately over the line, one frame for each direction in turn. Burst-mode activity has a repetition rate of 125 μs. This time is divided equally to enable a frame to be transmitted in each direction.

Transmission over the exchange connection is in fact half duplex. Buffers are therefore required in both transmitter and receiver because there is no longer continuous transmission in each direction.

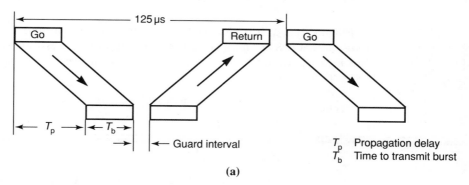

(a)

Figure 10.3 (a) Burst-mode operation.

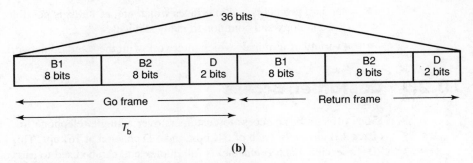

Figure 10.3 *continued*   (b) Burst mode cycle.

For continuous end-to-end transmission in each direction using burst-mode operation, the burst for each direction must operate at at least twice the end-to-end data rate, that is, 288 kbps. Unfortunately, because propagation time increases with line length, the instantaneous line rate must be further increased with distance. In practice, there are limits to the maximum transmission rate that simple telephone conductor pairs may support, hence burst mode is restricted to lines of a few kilometres.

EXAMPLE 10.1

Estimate the rate at which data must be transmitted over a line 4 km long using burst mode operation.

First, assume a typical velocity of propagation in the cable of 2×10^8 m/s. Therefore the propagation delay T_p is:

$$T_p = \frac{d}{v} = \frac{4000\,\text{m}}{2 \times 10^8\,\text{m/s}} = 20\,\mu\text{s}$$

From Figure 10.3(a) it is clear that the transmission time for one frame is 62.5 μs during which all 18 bits must be transmitted. Allowing for a propagation delay T_p of 20 μs, the time available to transmit the 18 bits contained in one burst T_b is only 42.5 μs. Therefore:

$$\text{Transmission rate} = \frac{\text{number of bits}}{T_b} = \frac{18\,\text{bits}}{42.5\,\mu\text{s}} \cong 424\,\text{kbps}$$

In practice, as indicated in Figure 10.3(a), a guard interval exists between consecutive bursts. This further reduces the time available to transmit a burst,

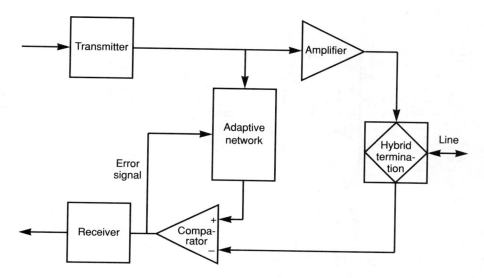

Figure 10.4   Echo-cancellation technique.

leading to an increase in transmission rate. Typical line signalling rates exceed 200 kbaud which is lower than Example 10.1 would suggest and made possible by the use of multilevel coding.

Alternatively, **echo cancellation**, shown in Figure 10.4, may be used where full-duplex transmission occurs on the exchange line. This is achieved by a hybrid termination at each end to combine and split the go and return signals. Transmission theory suggests that a line can support two signals propagating simultaneously in opposite directions without interfering with each other. However, the possibility exists for a station's received signal to be accompanied by echo due to its transmitted signal. Echo cancellation uses knowledge of the transmitted signal and appropriate subtraction from the received signal to cancel echoes. The technique is more expensive than burst mode because it requires sophisticated Digital Signal Processing (DSP) within the adaptive network. However, the line signalling rate is less than half that for burst mode operation and therefore longer lines may be supported so extending the geographical penetration of ISDN into the existing local-end network. Echo cancellation enables satisfactory operation on lines in excess of 10 km.

BT initially offered a pilot basic access which was only 1B+D and used burst mode transmission. Its current 2B+D service utilizes a proprietary local-end transmission protocol based upon echo cancellation. A three-symbol (ternary) line code is used, enabling the line signalling rate to be reduced to 240 kbaud. Nearly all local ends may support basic access, if required.

Burst mode and echo cancellation transmission are intended to connect customers to local exchanges via the local access network. Many PABXs in service today are fully ISDN based and use a four-wire connection throughout. Extension lines are of relatively short physical length. Therefore it is economically justifiable

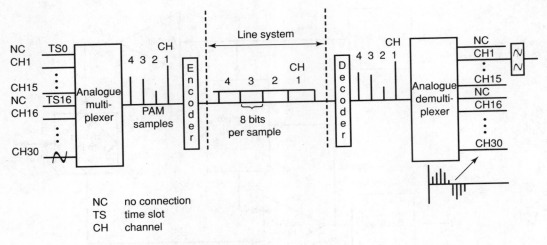

NC no connection
TS time slot
CH channel

Figure 10.5 PCM system.

to provide a separate pair for each direction of transmission so obviating the need for burst mode or echo cancellation techniques.

Alternatively, a primary rate service (or multiline access) aimed at PABX customers, based upon 2 Mbps digital line systems using 2 cable pairs, is offered, comprising up to 30 independent B channels and 1 D channel, and is known as 30B+D. The D channel, unlike basic access, is not available for data. Rather it operates at 64 kbps, the higher rate being necessary because it is the common signalling channel for all 30 B channels between line and exchange.

Figure 10.5 shows a 30-channel **Pulse Code Modulation** (PCM) system used between telephone exchanges to convey telephone connections. Each telephone or speech channel is sampled at 8 kHz to produce **Pulse Amplitude Modulated** (PAM) samples which are encoded into an 8-bit word, or pulse code. There are thus 256 different sample values or 'quantization' levels. Each 8-bit pulse code is time-division multiplexed together. There are a total of 32 time slots although in practice only 30 time slots are allocated for speech channels. The remaining 2 time slots are used for frame alignment at the receiving station and for signalling (common channel signalling), for example dial code information or clearing signals. Each channel's pulse code is transmitted serially over the line system, in turn. A complete system comprises sampling, encoding and multiplexing plus the converse functions for reception.

The line rate of a PCM system may be determined as follows:

$$\text{Line rate} = \text{sampling rate} \times \text{bits per sample} \times \text{no. of time slots}$$
$$= 8\,\text{kHz} \times 8 \times 32\,\text{bps}$$
$$= 2.048\,\text{Mbps}$$

PCM first appeared in the early 1960s as a 24-channel system. The United States, Canada and Japan continue to operate 24-channel systems but Europe now has 30 channels. ITU-T I-series recommendations have adapted the 30-channel PCM systems to provide primary ISDN access. Thirty 64 kbps PCM channels form the 30 B channels and the D channel uses one additional time slot. The remaining time slot is used for synchronizing the receiver clock with that of the transmitter by providing frame alignment between sending and receiving stations. Thus we see how the use of 64 kbps channels, although originated in PCM, has become a universal standard throughout the telecommunications industry.

Both Mercury's ISDN service and BT's primary rate access currently use the Digital Access Signalling System No.2 (DASS2) found in PCM systems for common channel signalling over the D channel. DASS2 is a forerunner to the international recommendation I.421 and is in fact very similar. In time DASS2 will be replaced by the internationally agreed standard I.421. DASS2 makes use of the HDLC protocol mentioned in Chapter 7.

Mercury does not offer basic ISDN services. BT's basic access signalling initially employed a system known as DASS1, from which DASS2 evolved. Ultimately, basic access systems will have to conform to the signalling requirements of I.420.

10.2.2 Customer installation

Having established the arrangements for provision of basic and primary ISDN access from the local exchange to the customer, we shall now consider how customers are able to interconnect their equipment to the PTT lines.

Figure 10.6 indicates the R, S, T and U **reference points** defined in the I-series recommendations for the interconnection of customer equipment to exchange lines. Most PTTs terminate the exchange line in the customer's premises using a Network Terminating Unit No. 1 (NT1). This unit is owned by the PTT and prevents customers from directly connecting their equipment to exchange lines at the U reference point. An NT1 for basic access provides services to the customer at the T reference point broadly equivalent to layer 1 of the OSI reference model, that is simple physical and electrical connection to the line. It also includes performance monitoring and maintenance (for example loop-back test facilities), bit timing, multiplexing of B and D channels and multidrop facilities.

Customer access at the T reference point, as has already been explained, conforms to the I.420 standard for basic access and I.421 for primary access. Within a customer's premises T and S buses are 8-wire passive buses. Two pairs are used for send and receive and the other four wires for the power feeding of terminals.

NT2s are optional devices for use with basic access only. They are intelligent devices and are available with various levels of functionality at layers 1 to 3 of the OSI reference model. An NT2 in its simplest form extends the I.420 interface from the T reference point to a number of S bus interfaces. Up to eight terminals may be multiplexed onto the S bus. An NT2 can provide switched voice and/or data facilities. Examples of NT2s are terminal controllers which offer multiplexing/ concentration, or PABXs which provide switching facilities. NT2s are also available which offer a LAN capability.

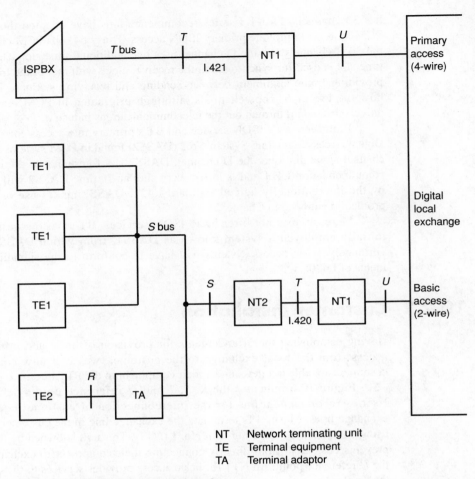

Figure 10.6 ISDN customer interfaces.

As indicated, ISPBXs perform NT2 functions and may be directly connected to a primary access via an NT1 at the *T* reference point. Non-ISDN PBXs which offer only plain old telephone services may also be connected simply to allow efficient use of an ISDN access and ultimately, in an all-ISDN scenario, to offer downward compatibility. However, to be realistic, by then non-ISDN customer equipment may well not be in existence.

A Terminating Equipment No.1 (TE1) is an ISDN compatible terminal, for example a videophone, and may be directly connected to the *S* bus using a standard RJ45 telephone connector. Provision also exists to connect non-ISDN terminals using a Terminal Adaptor (TA) providing a suitable adaptor is available. A maximum of eight TA and/or TEs may be connected to the bus. Non-ISDN terminals are called TE2s and are connected to a TA at the *R* reference point by one of a number of interfaces, for example V.24 or X.21.

10.3 ISDN applications

ISDN was originally conceived in the late 1970s. Early realizations in the 1980s resulted in a 'technology in search of an application'. At that time there was little requirement for anything more than simple POTS. The 1980s saw a change in this situation, first with the advent of PCs, and later as stand-alone PCs began to be replaced by networks of PCs. The demand for networking was in part addressed by using the existing telephone network in conjunction with a modem. The advent of ISDN offered the possibility of dispensing with modems and directly interfacing computer equipment into a network able to support at least 64 kbps and often much higher rates.

Apart from standard telephone provision, ISDN also offers a number of exciting new services:

(1) *Switched data*   Modems have been the traditional means of data transfer over earlier analogue telephone dial-up networks. Even today the highest data rate possible on a dial-up connection is 28.8 kbps with a modem. ISDN now offers a convenient switched facility for interconnecting computer and other digital equipment at a data rate of 64 kbps, or multiples thereof. Using ISDN for internetworking can be economic in many instances compared with a traditional approach which is based upon forming a computer network using leased lines.

Additionally, in time it will be possible for customers to connect to other networks via gateways. Currently it is possible to use an ISDN terminal to make a packet switched data call and much interest is now being shown in using the ISDN as a switchable LAN bearer.

(2) *Fax*   ISDN supports group 4 fax and, because of the relatively high data rate offered, faxes may be sent much more quickly than over the existing PSTN. Typically a full A4 sheet may be sent in a matter of seconds.

(3) *Teletex services*   These are a general-purpose alphanumeric messaging type of system. Apart from accessing Prestel, ISDN terminals, with suitable software, could be used for other similar services, typically home banking or shopping.

(4) *Photo Videotex*   This is one of the more novel facilities and is based upon the ability to transmit a full-colour photo in a matter of seconds. The classic application often cited is that of an estate agent. Details of houses for sale may be stored on a PC. This could include colour photos of the property including individual rooms and garden aspects. By means of an ISDN call, such text and visual information may be accessed remotely. This enhancement of the traditional postal circulation of house details is much more informative and may well reduce the number of fruitless journeys which prospective purchasers often make to view properties for sale. Photo Videotex has enormous possibilities, both in private advertising, such as second-hand cars, as well as an alternative approach for mail order and other forms of direct selling.

(5) *Slow-scan TV* A normal domestic TV uses 625 picture lines and requires appreciable bandwidth to produce moving images. Still images can be produced by increasing the time for each line scan and thus reduce the bandwidth necessary. The down side is that fast-moving images result in poorer quality. ISDN offers slow-scan TV using a single 64 kbps channel by means of a series of images which are gradually overwritten, line by line. A complete picture change may occur in 1 second. Applications are the connection of remote video surveillance camera signals to a central monitoring site, or remote monitoring of hazardous areas.

(6) *Videophone* With recent advances in the capabilities and speed of signal processing techniques, by means of data compression techniques using DSP, real-time TV signals of adequate quality may be transmitted over a basic access connection. This has resulted in the recent launch of a commercially available videophone.

(7) *Videoconferencing* An extension of the videophone is videoconferencing where three or more parties, each with a videophone, may be interconnected. Videoconferencing using an ISDN connection employs the same basic technology as a videophone. Typically, a multimedia PC and camera are required. In addition, the technology to support multiple connections is necessary and is readily available within ISDN.

(8) *Fast file-transfer* As already indicated, a dial-up connection may operate at speeds up to 28.8 kbps over the PSTN, although such high rates cannot be guaranteed on every connection. ISDN offers multiples of 64 kbps, enabling reliable fast file-transfer between, say, a pair of PCs.

In conclusion, a modern ISDN installation is a multifunction terminal which is now available in the marketplace. It comprises a normal telephone handset, screen and keyboard plus a printing facility. Such a terminal can support all of the above facilities.

10.4 ISDN operation

We have discussed the general concept of ISDN, examined its underpinning technology and considered a range of services that it may support. Now we shall consider the operation of ISDN.

The following modes of operation are supported:

- circuit switched connections
- semi-permanent connections
- packet switched connections
- frame relay
- frame connection

A circuit switched connection requires a circuit, or virtual path, between the two users via the network. This is achieved using control signals over the D channel to the local exchange using layers 1, 2 and 3 of the OSI reference model. The telephone network uses these signals to set up suitable connections to route the call to the intended destination. It does this using a separate signalling network which operates in accordance with ITU-T Q-series switching and signalling systems recommendations for ISDN. One of the key elements of these recommendations is known as Common Channel Signalling No. 7 (CCS7). As its name suggests it is based upon common channel signalling principles and is for use in the various switches within the network (compared with BASS which is used for user signalling). The standards conform to OSI principles and make use of all seven layers of the reference model. Upon establishment of a circuit switched connection, each B channel offers a transparent 64 kbps transmission path which the user may use for any purpose, for example, one speech path or a number of lower speed data streams multiplexed together.

Semi-permanent connections are an extension of circuit switched connections. They may be private circuits, in which case a permanent connection is established between two users. Alternatively, infrequent connections may be provided by the PTT operator for a period of time in accordance with a predetermined pattern, for example, two hours one day per week. Because the connection is already established when it is required, only layer 1 functionality is necessary.

Packet switched connections are supported using a three-layer protocol similar to X.25 (see Chapter 8). However, many public data networks now offer a gateway to interconnect ISDN traffic. Interconnection arrangements for packet switched ISDN connections via the PDN are defined in the X.31 and I.462 recommendations. Initially, a circuit connection is established between user and gateway as for a normal circuit switched call. Having established a connection to the PDN gateway a 64 kbps circuit now exists over which data packets may flow. In order to make packet switched connections through the PDN the user's packet-mode terminal uses the X.25 protocol at layer 3 whereas layers 1 and 2 use I.420/421. As an alternative, X.25 packets may also be routed through the ISDN to the gateway over the D channel.

The concept of frame relay and its relative merits compared with packet switching have already been discussed in Chapter 8. ISDN supports a frame relay service which is defined in the I.122/Q.922 recommendations. It allows several calls, each to different locations, to exist simultaneously. This is achieved by assigning each call a virtual path. Frame relay is a connectionless protocol. ISDN is also able to support connection-oriented operation, that is frame switching. However, frame relay dominates and is also being used widely in private networks. Both services use the same signalling procedures but frame switching requires that the network supports error and flow control.

10.5 Availability

Finally, having established what ISDN is, its capabilities and technologies, let us consider its potential in terms of current availability. In the United Kingdom

Mercury provides primary access in the form of its 2100 'Premier' service aimed at medium to large business customers. The service is available to about 80% of the business community in areas where Mercury has installed its own cables, or to any customer able to access one of its microwave radio sites. In the latter case customers are not normally more than 35 miles from a site. Service is therefore available in nearly every town and city.

BT can readily offer ISDN to customers connected to digital exchanges. Currently 80% of business customers have such access. It markets two ISDN services. ISDN-2, which is a basic (or 2B+D) service, and ISDN-30, which is a primary (or 30B+D) service.

Within the United Kingdom, and indeed internationally, there exists a mixture of ISDN and the much more prolific PSTN customers. Where an ISDN call is originated, be it national or international, it is routed entirely via ISDN exchanges, if possible. Otherwise the call is set up in part via a PSTN and hence service drops back to that of a normal PSTN call. This means that only simple telephone facilities are then possible over the connection.

Although most EU nations are currently building ISDN networks as well as countries such as Japan, the United States and Australia, no fully international service as yet exists. BT offers international ISDN, both at basic and primary rate, to 18 countries including most of Europe, North America, Australasia, Hong Kong, Singapore and Japan. Many terminals in the United States operate at 56 kbps for which rate adaption is additionally required.

Mercury's international access is able to support fully primary access on all routes. It has some international routes of its own and also uses BT's via its interconnectivity right. In consequence, international coverage is similar to that of BT with access to 22 countries.

Exercises

10.1 Define the terms IDN, IDA and ISDN.

10.2 Explain the differences between a PSTN and ISDN.

10.3 Discuss, with reasons, the services which ISDN is able to support. Contrast them with their provision, if possible, over a conventional PSTN line.

10.4 A company is considering dial-up access to a remote computer using the PSTN in conjunction with a modem, or installing an ISDN line. Compare their relative merits.

10.5 Discuss the functions of the following: NT1, NT2, TE1, TE2 and TA.

10.6 An NT is effectively the interface between the customer's installation and the PTT's domain. Discuss what functions an NT is required to offer.

10.7   Compare the relative merits of burst mode and echo cancellation techniques used in ISDN local ends.

10.8   State the line transmission rate of a 2B+D line which uses echo cancellation. Assume binary data signals.

10.9   Estimate the rate at which data must be transmitted over a line 6 km long using burst mode operation assuming 2B+D transmission.

10.10 A 2B+D line is to use burst mode transmission. If the maximum signalling rate is not to exceed 500 kbps (assuming binary coding), calculate the maximum length of line which may be supported.

Broadband networks

Broadband ISDN (B-ISDN) is a further development from ISDN which is now starting to appear. It has the potential to support services that require transmission rates far in excess of those for basic telephony. ISDN (see Chapter 10) was the forerunner of B-ISDN and was only originally conceived to provide high-quality telephony services using improved digital technology. In addition, ISDN was able to offer what at that time was high-speed data transmission on a circuit switched basis. There was no requirement originally, as now, to accommodate video communication, such as video-mail, and multimedia applications, such as cooperative computer working via voice and video accompanied by high speed-file transfers, which B-ISDN is able to support.

This chapter indicates the type of services B-ISDN may support, broadly based upon the former CCITT recommendations. Consideration of the needs of image communication is discussed, with its attendant high data rate and large dynamic range. Finally, Asynchronous Transfer Mode (ATM) is introduced. ATM is a bit transportation mechanism which is highly versatile and adaptable and currently under development to support B-ISDN. It can efficiently transmit current and envisaged services, irrespective of bit rate, be they speech, data or video, and is currently an extremely active R & D area in data communications. ATM is set to transform WANs in the next century because it offers the potential to replace and subsume all existing WANs and can support all future services. The section on ATM outlines its basic operation, considers its relationship with existing technologies and discusses the emerging standards.

11.1 Broadband ISDN

The ITU-T transmission rate classifications for B-ISDN and ISDN or narrowband ISDN (N-ISDN) are compared in Table 11.1.

B-ISDN supports much higher bit rates than ISDN. Indeed, data rates of 600 Mbps and beyond are also under consideration and will become commonplace early in the 21st century. The distinction between ISDN and B-ISDN is not simply an arbitrary one. The three B-ISDN rates shown are easily derived within the current Time-Division Multiplexing (TDM) arrangements based upon 2 Mbps PCM

Table 11.1 Transmission rate classifications.

Narrowband		Broadband	
D	16 or 64 kbps	H_2	30–45 Mbps
B	64 kbps	H_3	60–68 Mbps
H_0	384 kbps	H_4	120–140 Mbps
H_{11}	1.536 Mbps		
H_{12}	1.92 Mbps		

systems which are shown in Figure 11.1. Four PCM channels multiplexed together produce level 1, the primary level, of the trunk hierarchy which is 8 Mbps. Further multiplexing yields the secondary level at 34 Mbps, and thereafter 2 or 4 channels may be further multiplexed. Thus it may be seen that B-ISDN rates closely match, or may be derived from, rates available within existing TDM hierarchies.

With the exception of image communication, current voice and data requirements are largely catered for by existing network technologies. Image communication has evolved relatively slowly since the first terrestrial broadcast TV systems appeared 40 or 50 years ago, largely being limited by technology. The concept of a videophone, which is a relatively simple progression from a purely voice-based network, on a public switched basis in a similar manner to that of a normal telephone system has for instance only recently become a commercial reality. It is predominantly the emerging and proposed video applications, for example videomail, that are driving the long-term development of B-ISDN networks. Such networks offer network operators the ability to combine separate voice and data networks into a single network encompassing local end, switching and transmission and hence make efficiency gains.

Both ISDN and B-ISDN services need to be delivered right into the customer's premises. As already seen in Chapter 10, traditional metallic conductors installed in the local-end network can support ISDN with its attendant data rate of up to 2 Mbps. However B-ISDN, with its relatively high transmission rates, has until recently only been found within the more hospitable environment of a PTT's domain. Some lower speed B-ISDN services demand speeds only of the order of tens of Mbps and may be supported by metallic transmission media. However,

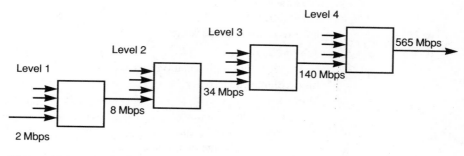

Figure 11.1 TDM hierarchy.

higher rates of the order of hundreds of Mbps will only be possible for customers using fibre technology. The delivery of such rates for broadband services to customers is in its infancy because these rates are not yet widely available.

11.1.1 B-ISDN services

The number of possible services, both now and envisaged, is dependent only upon one's imagination! Services may be classified as *interactive* where information is two-way and *distribution* where information is primarily one-way and which may, or may not, allow a user presentation control.

Services may also be categorized as: *conversational*, such as a telephone connection; *messaging*, where there is no need for real-time activity between communicating parties and includes X.400 e-mail and Teletex; *retrieval services*, where a user may select information; *distribution* (or broadcast) services, which provide a continuous flow of information to a user. Such services may or may not offer user presentation control. A distribution service with user presentation cycles information repetitively enabling a user to select the start of or to reorder presentation. An example of such a service with user presentation control is Teletex. An example of where there is no user presentation control is that of national TV distribution.

Interactive services

The following selection of interactive services represents something of the scope or range which may be reasonably envisaged. The list has been classified to some extent in order of bit rate, slowest first.

- *Message services*
  - document mail
  - video mail
- *Retrieval services*
  - data retrieval
  - document retrieval
  - broadband videotex
  - video retrieval
  - high-resolution image retrieval
- *Conversational services*
  - high-speed digital information services
  - high-volume file transfer
  - high-speed telefax
  - broadband videotelephony
  - broadband videoconference
  - high-resolution image communication

Distribution services

The following list of possible distribution services is again ranked in approximate order of bit rate:

- Existing TV distribution
- Pay TV
- Document distribution, for example electronic newspaper
- Full-channel broadcast videography (text, graphics, sound, still images)
- Video information distribution
- Extended-quality TV distribution
- HDTV distribution

11.2 Image communication

Consider the general requirements of image communication with reference to Figure 11.2. A digital signal may be considered as a succession of words, each of A bits, representing a single pixel. Frames consist of $B \times C$ such samples and the frame rate is D frames per second. This results in a total bit rate for image communication which is the product of A, B, C and D. Even with modest resolution and quality, the bit rate of image signals becomes relatively excessive at several hundred Mbps. (Note that if resolution is doubled the number of samples, and hence bit rate, is quadrupled.) Table 11.2 shows existing and proposed bit rates that require support.

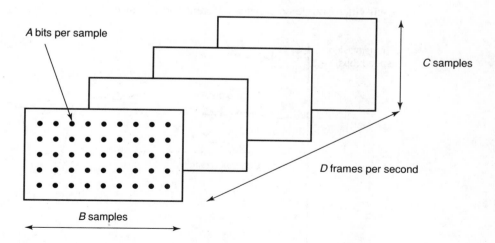

Figure 11.2 Image bit rate requirements.

Table 11.2 Current and envisaged bit rates.

System	Bit rate (Mbps)
Videophone	0.2
Videoconference	24
Digital Broadcast TV	83
HDTV	883
SHDTV	1600–40 000 (under discussion)

Traditionally, many channels have been assigned to image signals on a **Constant Bit Rate** (CBR) basis, that is, the channel operates at a single transmission speed which is available throughout the connection. From an information point of view this may mean that the channel represents overprovision when transmitting stills and slowly moving images. Where action is rapid, the channel may not offer sufficient capacity and distortion may occur at the receiver. One way of looking at this is shown in Figure 11.3 where channel capacity is static but picture quality varies over time. Ideally, picture quality should remain constant irrespective of changes in the image but this implies that the bit rate must now vary. For efficient and economic operation what is required is a channel which can match the instantaneous bit rate required and of course the maximum for a specified quality. Figure 11.4 shows such a **Variable Bit Rate** (VBR) arrangement. Note the 'symmetry' with Figure 11.3.

Video signals, as indicated above, have a high ratio of peak to mean bit rate. If channel capacity seeks to support the peak demand it may be fully utilized only for a very small proportion of the time. For much of the time it represents an overprovision in channel capacity.

Consider N video sources, each of average bit rate R, sharing a network. If each video source conveys a different signal (for example each one is a different

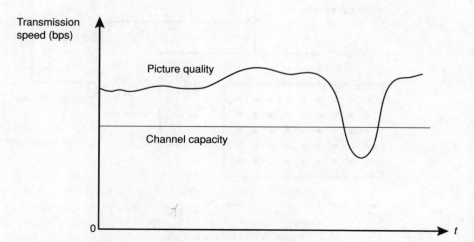

Figure 11.3 CBR transmission.

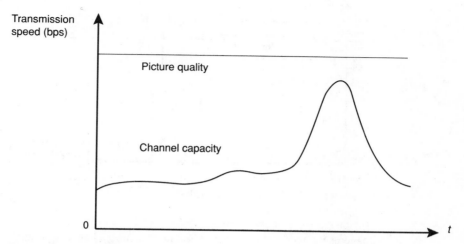

Figure 11.4 VBR transmission.

channel or network broadcast) they may then be regarded in statistical terms as uncorrelated, or independent. Hence, irrespective of their peak bit rates, the average load presented to the network is given by:

Average load of N independent sources $= N \times R$ bps

Utilization U is defined as the ratio of average load to network capacity C:

$$\text{Utilization} = \frac{\text{average load}}{\text{network capacity}} = \frac{NR}{C}$$

Suppose that the network design considered only economic efficiency. Then network capacity C should be made equal to, or slightly greater than, NR to yield a utilization of about 1. Statistically, however, the sources will from time to time aggregate to a load in excess of NR. Under these conditions overload and loss of signal (drop out) results. There are a number of approaches to the solution of this problem. One is to increase the network capacity C, which increases the cost of the network. From a user perspective, it increases the quality of service offered. In addition, the time for which the network is underutilized is also increased. Such a solution is unattractive to PTTs. The main alternative is to use some form of coding of information signals to reduce their mean bit rate. Data compression is one such technique. Two-layer coding for video is another technique and is currently the subject of much research and appears to be the way ahead for image communication.

A two-layer coder for video is shown in Figure 11.5. Once satisfactory communication is taking place, most packets are **guaranteed packets** which are produced by the first-layer coder. Very few **enhancement packets** then occur. The first-layer coder generates essential packets which are given priority and guaranteed transmission by the network, for example synchronization signals. Guaranteed packets also contain low-quality picture information essential to maintain a tolerable

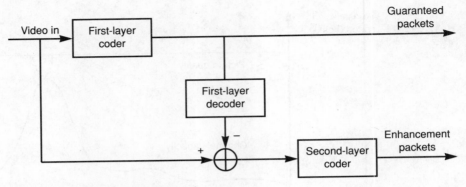

Figure 11.5 Two-layer video coder.

image at the receiver at all times. Enhancement packets are generated only when a change occurs in the picture, for example in colour or movement. If the image is constant, the output after first-layer coding and decoding is the same as the current video signal. In consequence no output occurs from the summer and thus no second-layer packets are produced. Such packets are in fact only generated when a change occurs in the picture. Second-layer packets enhance picture quality but are not guaranteed transmission. If enhancement packets are not transmitted, picture change would, over time appear, at the receiver. The transitions between changes would simply be of inferior quality.

If the network is approaching congestion, it can reduce the load presented to it by accepting only guaranteed packets. Under such circumstances which it must be stressed may occur only very infrequently and then only for a short duration, users suffer some impairment. However, guaranteed packets ensure an acceptable quality. Two-layer coding therefore provides a good balance between quality and bandwidth and is an example of the use of VBR transmission.

For widespread image communication CBR is uneconomic and a new transport mechanism is currently the subject of much discussion and development. This transport mechanism which offers VBR is based upon packet-switching principles and is discussed in the following section.

11.3 Asynchronous transfer mode

Broadband systems of the future are unlikely to use the switching techniques currently employed within WANs, largely because they will not offer the capacities demanded of emerging B-ISDN services. **Asynchronous Transfer Mode** (ATM) is the emerging switching technology which will support B-ISDN and is a high-speed packet-based network which can transport any type of service, be it voice, data or video. The aim of ATM is to provide a single multimedia broadband network to support all current and envisaged services, that is, from telemetry through voice and data to the very high speed requirements of video services such as SHDTV and

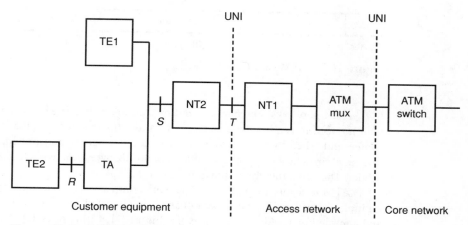

Figure 11.6   B-ISDN network topology.

virtual reality. We are currently in a transitional stage where some of these services are in their infancy and are not yet widely available. However, there are already in existence a number of broadband services which may benefit from ATM today. These services include frame relay, MANs and **Switched Multimegabit Data Services** (SMDS).

Frame relay and MANs were discussed in Chapters 8 and 9, respectively. SMDS is currently perceived as a LAN interconnect facility and therefore may be regarded as part of broadband technology. It offers a high-speed connectionless packet-switching facility for customer access to a WAN at multimegabit speeds of 2, 4, 10 or 25 Mbps.

B-ISDN topology at the customer's premises, which is similar to ISDN (see Chapter 10), is shown in Figure 11.6. Within the customer's site TE1, TE2 and TA elements are defined and perform in a similar manner as for ISDN. The boundary between the customer's equipment and the public network is, as with ISDN, at the T reference point. This boundary between user and network is also known as the **User–Network Interface** (UNI). The access network connects customers, via an NT1, to the local exchange. This network may be owned and operated by a PTT or by some other operator such as a cable TV provider. The core network contains the transmission and switching elements of the ATM network and is the province of a PTT. The interface between the access and core networks is known as the **Network Node Interface** (NNI).

Packets which are of a fixed length of 53 bytes are called **cells**. Cells generally originate at the customer's premises intermittently. To transport ATM cells efficiently over the core network, they must be presented to the ATM switch in the core network (which is in effect the local exchange) as a continuous stream. There are a number of options within the access network to connect customers to an ATM switch. Customers who generate sufficient traffic may be directly connected to an ATM switch. Otherwise, a number of low-traffic customers are normally connected to an switch ATM multiplexer which is in turn connected to a switch.

Header	Information
5 bytes	48 bytes

Figure 11.7 ATM cell.

ATM networks are characterized by stations generating cells as required and with no need to send if there is no information. Information is initially buffered at source and when there is sufficient information a cell is transmitted; hence cells originate asynchronously. ATM cells contain 48 bytes for user data and a 5-byte header. The cell structure is shown in Figure 11.7.

ATM transmission in the core network contrasts with traditional time-division multiplexed networks where each channel is allocated a fixed slice of bandwidth and sampled at regular time intervals. Figure 11.8 illustrates CBR transmission which is also known as **Synchronous Transfer Mode** (STM). If a particular input to an STM multiplexer has little or no information to send, it still has its slice of bandwidth continuously available to it because the number of time slots allocated to each input is constant and they occur repetitively. STM networks can handle high-volume trunk telephony and traditional broadcast TV quite satisfactorily but are unsuited to the variable and high bit rate requirements of B-ISDN which ATM networks are planned to support. Such services may be very bursty and are anticipated to have raw bit rates of the order of 10^7–10^{11} bps. ATM transmission, also shown in Figure 11.8, produces fixed-length cells from each input. Cells are then transmitted on a more or less first come, first served basis and hence occur asynchronously. ATM may arguably be placed between circuit switched networks characterized by STM and packet switched networks where packet length is variable. ATM is in fact analogous to statistical multiplexing and is a compromise between the needs of CBR services and those of bursty packet-based services.

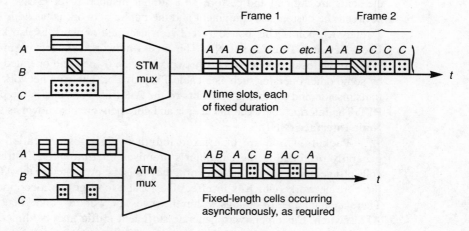

Figure 11.8 STM and ATM transmission.

Table 11.3 PDH standards (all rates in Mbps).

	United States	*Japan*	*Europe*
Level 1	1.544	1.544	2.048
Level 2	6.312	6.312	8.448
Level 3	44.736	32.064	34.368
Level 4	139.264	97.728	139.264

The transmission element of core networks currently employ **Plesiochronous Digital Hierarchy** (PDH) which was shown earlier in Figure 11.1. PDH is based on multiplexing four 2 Mbps PCM tributaries to form an 8 Mbps stream which in turn may be multiplexed to form 34 Mbps, and so on. Unfortunately PDH standards vary throughout the world, as Table 11.3 shows.

Interworking internationally at a particular level using PDH requires special arrangements where rates are not the same. A detailed discussion of multiplexing is beyond the scope of this text but, just as with 2 Mbps PCM systems discussed in Chapter 10, each level of multiplexing requires additional overheads, for example frame words. The signals' structure at any one level, especially higher levels of multiplexing, are complex, being a compounded version of lower level structures. As a result, 2 Mbps (or indeed 64 kbps) tributaries cannot readily be inserted or extracted from level 2 streams and above. Rather, to perform such an operation full multiplexing, or demultiplexing, between a given level and 2 Mbps must be performed. In an ideal core network, as envisaged with ATM, primary tributaries should be able to be inserted or extracted at all levels of the multiplex hierarchy to provide flexible routing and efficient use of the network.

In 1986 ANSI developed a transmission standard using standard frame formats and signalling protocols which overcame the above difficulties associated with PDH. The standard is the **Synchronous Optical Network** (SONET). The CCITT, and more latterly ETSI, built upon the concepts of SONET, which has led to the development of a **Synchronous Digital Hierarchy** (SDH) standard for global use. SONET applies to optical transmission systems operating at 155 Mbps and hierarchies to 622 Mbps and 2488 Mbps and beyond. ATM has adopted SDH at 155 Mbps as its primary transmission rate. However, accepting that many PTTs have a long way to go in converting their core networks from PDH to SDH, and that this rate cannot yet be widely delivered to customer premises, lower interim rates such as 34 Mbps will have to be used.

ATM is able to support the virtual channels (VCs) already described in Chapter 8 as well as **virtual paths** (VPs). Figure 11.9 shows the relationship between VCs and VPs and how a number of them in combination may be transmitted over the same transmission 'pipe' or link. A VP simply consists of a number of VCs, each of which has the *same* end points. To switch VPs and VCs correctly through the ATM network, each VP and VC is allocated an identifier, that is a VPI and VCI, respectively.

The ATM cell header fields used for cells at the UNI are shown in detail in Figure 11.10(a). The NNI is the interface between ATM switches within the core

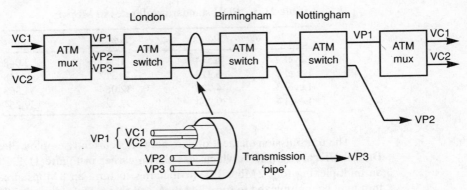

Figure 11.9 Virtual channels and virtual paths.

network. The ATM cell header fields for cells at the NNI are almost identical to that for the UNI and are shown in Figure 11.10(b). The fields are used as follows:

- **Generic Flow Control** (GFC) Manages multiple terminals accessing the same connection. As yet the exact mechanism for control has not been defined.

- **Virtual Path Identifier** (VPI) Used to route ATM cells over a VP. Where a connection may be routed over a number of different VPs, the field's value will change as a cell is switched to a different VP.

- **Virtual Channel Identifier** (VCI) Similar to VPI but used for VCs.

GFC		VPI	
VPI		VCI	
VCI			
VCI		PT	CLP
HEC			

(a)

VPI			
VPI		VCI	
VCI			
VCI		PT	CLP
HEC			

(b)

Figure 11.10 ATM cell header field: (a) UNI, (b) NNI.

- **Payload Type** (PT) Indicates general type of data in cell, for example maintenance, last cell of a multicell message.

- **Cell Loss Priority** (CLP) There is always a small risk that a cell may be lost as it passes through a network. This may be due to congestion or overload. The CLP field is used to indicate to the network whether or not the cell may be discarded in favour of another cell. For example, the CLP bit can distinguish between the guaranteed and enhancement cells discussed in two-layer video coding in Section 11.2.

- **Header Error Check** (HEC) Performs a cell header checksum over all fields, except GFC if present, for error control. It may also be used to give an indication of the start/end of cells in a stream of cells appearing at a node.

Now let us consider switching within the context of ATM. Switches consist of a number of input and output ports and any input may be switched to any output. Each port carries a stream of ATM cells operating at 155 Mbps. This stream may not pass directly into the switch. Often cells are stored in an input buffer and then passed to the output port once the switching path has been established. Commonly an ATM switch is a 64 × 64 port device and is therefore required to handle throughputs of the order of tens or hundreds of Gbps. At such speeds switching must be performed in hardware. Switching itself is performed by examining the cell header and connecting an appropriate switching point to interconnect to the required output port.

The time taken for a cell to reach a destination varies. Cell delay occurs for several reasons:

- transmission path length
- the time taken to fill a cell at the source
- delays associated with input buffers within switches

Cell delay, unless excessive, need not be of concern providing that every cell suffers the same delay. In this case cells are originated and received at the same rate. What is important is cell delay variation. There is little scope for control of the first two causes of cell delay. Queuing delay occurring at input buffers to switches may vary beyond what is desirable if the network is not well ordered. Buffers are generally dealt with on a FIFO basis as explained in Chapter 8. Under heavy load cells may enter a buffer at a faster rate than they are retrieved. This means that the delay between storing and retrieving a cell progressively increases, resulting in variable cell delay. This situation cannot be sustained for very long since in extreme cases demand for buffer space may be exceeded, in which case cells are actually lost.

11.4 ATM standards

Some standards for ATM have begun to appear, initially via the CCITT and now the ITU-T sector. Most standards are only provisional and subject to further study. The

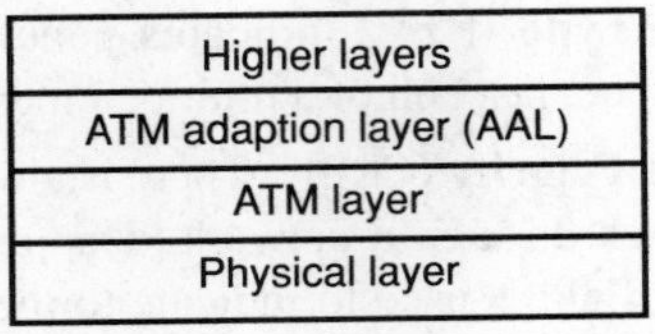

| Higher layers |
| ATM adaption layer (AAL) |
| ATM layer |
| Physical layer |

Figure 11.11 B-ISDN protocol reference model.

standards necessary for ATM operation are encompassed within the I-series recommendations for B-ISDN.

The I-series recommendations which specifically relate to ATM are I.150, I.361, I.362 and I.363. The I.150 recommendation defines ATM functional characteristics and deals with multiplexing of cells, switching (using packet switching techniques), quality of service of cell transmission, payload types and identifier field, and user information flow control.

I.361 defines the ATM layer specification. The recommendation simply specifies the ATM cell structure and which has already been discussed. The ATM layer is found in the B-ISDN **Protocol Reference Model** (PRM) which reflects the OSI reference model. Part of the PRM is shown in Figure 11.11.

The I.362 recommendation defines the **ATM Adaption Layer** (AAL) functional description. The AAL adapts in the sense that the layer may offer one of four possible classes of service as listed in Table 11.4.

There are three characteristics which distinguish class of service:

(1) constant/variable bit rate

(2) connectionless/connection-oriented mode

(3) requirement, or otherwise, of a timing relationship between source and destination

The first two characteristics are self-explanatory. The need for a timing relationship between source and destination depends upon the type of service. Class A is suitable for CBR video and circuit emulation of ISDN, that is $n \times 64$ kbps connections up to a maximum of 2 Mbps. Class B provides VBR voice and video services. Both of these classes of service require a careful timing relationship between source and destination to ensure that delay in individual cell transmission time through the ATM network remains appreciably constant. This ensures that the

Table 11.4 AAL classes of service.

Class of service	Example of service
A	Circuit emulation
B	VBR – video and audio
C	Connection-oriented data transfer
D	Connectionless data transfer

Table 11.5   Class of service characteristics.

	Class A	Class B	Class C	Class D
Bit rate	Constant	Variable	Variable	Variable
Connection mode	Connection-oriented	Connection oriented	Connection oriented	Connectionless
Timing relationship required	Yes	Yes	No	No

sampled data that is being transmitted, and typical of the services being offered, does not suffer unacceptable jitter at the destination. Class C is suitable for VBR services such as packet switched and frame relay services. Class D is able to support a LAN interconnect facility.

Table 11.5 compares the characteristics of each class of service.

The I.363 recommendation deals with the AAL specification which describes the interaction between the AAL layer, the layer above and the ATM layer below. The recommendation additionally describes the AAL peer-to-peer operation.

Signalling standards are still under development. The ITU-T has defined a phased roll-out of B-ISDN signalling protocols:

- *Release 1*   Q.2931 is being defined for user access signalling and is an extension of Q.931 used for N-ISDN. Network signalling is to use the user part of CCS7 and is to be known as B-ISUP. When approved, network signalling standards will become Q.2761-4 recommendations. These protocols will support point-to-point communication.

- *Release 2*   This, and release 3, have not yet been agreed. Release 2 will define VBR standards for multipoint and multiconnection operation. It will not offer more advanced multimedia applications or TV-based services.

- *Release 3*   This will complete the full roll-out of B-ISDN signalling protocols to support the full range of services including multimedia and distributed services.

In conclusion, ATM is capable of flexible and adaptable operation able to meet all perceived needs of future services. It holds the prospect of a general-purpose transport mechanism able to carry fully integrated and high-speed services. The supporting technology for ATM has already been developed although its availability is patchy. ATM enjoys strong support within the standards arena and in consequence appears to be in an excellent position to be adopted as a next-generation network to replace the WANs currently offered by PTTs. ATM is also under consideration as a much improved replacement for traditional LANs. With the speed and performance demands of such networks increasing, ATM may also replace computer networks generally in the next century.

Exercises

11.1 Define the terms ISDN and B-ISDN and draw a distinction between them.

11.2 Discuss the reasons for the emergence of B-ISDN.

11.3 Discuss the services which B-ISDN has the potential to offer.

11.4 Define the term 'Asynchronous transfer mode'.

11.5 Distinguish between CBR and VBR transmission.

11.6 Explain how ATM may support both CBR and VBR transmissions.

11.7 Explain clearly why constant bit rate networks are unsuitable for communication of image signals with high peak to mean bit rates.

11.8 Coding is seen as a way of reducing the capacity that a network is required to offer its customers and yet still provide adequate transmission quality. Discuss how such coding enables these two conflicting parameters to be reconciled.

11.9 Discuss the use of the cell loss priority field contained within the header of an ATM cell.

11.10 Some ATM connections require a timing relationship between connected stations. Explain what this means and why it might be necessary.

11.11 Outline the scope of the following standards: I.150, I.361, I.362 and I.363.

Network management

12

A data communications network needs to operate efficiently, particularly in the event of a major failure. The cost of a network ceasing to function as a result of a failure is extremely high and most major networks incorporate some form of network management system to ensure its efficient operation. The functions performed by a network management system must include the monitoring of the network's performance, handling failures when they occur and reconfiguring the network as a response to such failures.

12.1 Performance management

A number of measures of network performance were mentioned in Chapter 6 in relation to link control and management. These included link throughput and utilization, error rates and delay times. These measures apply equally to the network as a whole and the collection of these and other statistics forms an important part of any network management system. Statistics gathered by a network management system not only aid the efficient operation of a network but can also be used to plan the future requirements of a network. Most systems use the two further performance criteria of availability and response time.

12.1.1 Availability

In most communications systems availability is an important measure of the amount of time that a system is available for use. Typically it is measured as the average amount of time a system is available per unit time. Alternatively it can be measured in terms of **Mean Time Between Failures** (MTBF) and the **Mean Time To Repair** (MTTR) as follows:

$$\text{Availability} = \frac{\text{MTBF}}{\text{MTBF} + \text{MTTR}}$$

Typical values for MTBF and MTBR for various network components are shown in Table 12.1.

Table 12.1 Typical MTBF and MTTR values.

Network component	MTBF (hours)	MTTR (hours)
Processor	200	2
Modems	5000	3
Lines	3000	4
Terminals	1000	2

EXAMPLE 12.1

A processor has a mean time between failures of 200 hours and a mean time to repair of 3 hours. Determine the processor's availability.

$$\text{Availability} = \frac{\text{MTBF}}{\text{MTBF} + \text{MTTR}}$$

$$= 200/203 = 98.5\%$$

12.1.2 Response time

In many communications systems response time is an important measure of performance. It is a measure of the speed of operation of the system. It can be

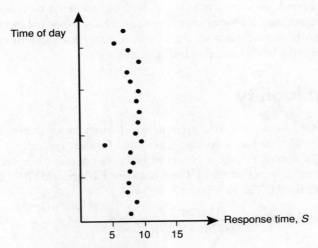

Figure 12.1 Response-time chart.

Table 12.2 Response times as percentages.

Time (s)	Data transfers (%)
0–1	0
1–2	0
2–3	0
3–4	3
4–5	6
5–6	13
6–7	27
7–8	37
8–9	10
9–10	3
10–11	0
11–12	0

defined for systems in which humans carry out operations as the time between an operator pressing a transmit key and a reply appearing on a screen. This time is also known as round-trip delay. A network management system will gather response time statistics for specific devices and circuits as well as for complete networks. Once gathered, statistics are presented in a variety of ways, quite often in the form of a graphical printout. Figure 12.1 shows a typical response-time chart.

This chart shows response times plotted at regular intervals, the value plotted being an average response time obtained during the time interval. Network management systems often use such statistics to produce more detailed breakdowns which allow percentiles to be obtained, as in Table 12.2. which has been obtained from Figure 12.1.

12.2 Failure management

An important feature of any network management system is the detection of failures and the subsequent repair and restoration of system performance. The first step in failure handling following the detection of a failure is normally setting off an alarm of some kind. Traditionally, data communications equipment such as modems and multiplexors have used a crude form of alarm in which the loss of a circuit causes a lamp to extinguish. This negative type of alarm conveys little information and can easily be ignored. A true network management system normally provides a more positive type of alarm such as a message at an operator's desk indicating the location and type of failure which requires some action by the operator. Some systems also set off alarms to indicate degrading situations, which may allow a failure to be averted.

Once an alarm has warned of a failure the next step is to restore the system to its normal operating condition. Initially, this will probably involve finding some short-term solution to the problem such as a fallback arrangement, but eventually the precise cause of the fault will need to be determined by using test equipment and the faulty device repaired or replaced.

12.2.1 System restoration and reconfiguration

System restoration following a failure involves two processes. Normally, the first step is some form of **fallback switching** which involves the replacement of a failed device or circuit by an immediately available backup. In the case of failed circuits the backup is normally the PSTN or, more recently, the ISDN. With equipment such as modems, terminals and multiplexers, the backup equipment is usually taken from a readily available pool of spare equipment. Fallback switching is normally automatic but may still be carried out manually, for example by using a **patch panel** to reconnect alternative circuits temporarily. A patch panel is an arrangement that allows devices to be connected, or patched, together. Its principle of operation is similar to an old-fashioned manual telephone switchboard in which telephone circuits can be switched by connecting circuits into sockets on the switchboard. Thus, a patch panel may be used to bypass faulty equipment and to patch in spare equipment in the event of failure. Similarly, a patch panel allows a limited amount of reconfiguration to be carried out, although the reconfiguration of a large network is normally beyond its capabilities.

12.2.2 Test equipment

Fallback switching is only a short-term solution to network failure. The more permanent repair or replacement of equipment or circuits requires testing to locate a failure precisely. Many systems provide an integrated test facility although some may only provide access for separate test equipment. Three items of test equipment have traditionally proved useful, particularly in smaller networks in which there is no integrated test facility.

Breakout box

This is the simplest item of test equipment which allows access to V.24 interfaces. The breakout box is inserted into the interface between the DCE and DTE. The status of the V.24 lines are indicated by lamps and switches allow an operator to simulate the operation of the interface.

Data analyser

A step up in sophistication from a breakout box is the data analyser. This is a microprocessor-controlled device which can capture the status of data on a data line

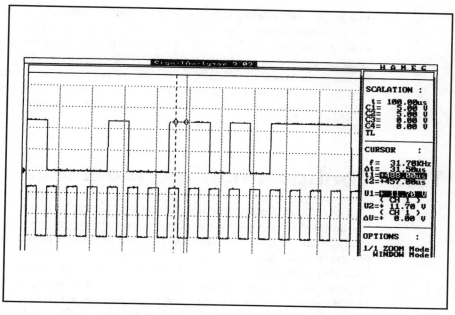

Figure 12.2 Data analyser printout.

and display this information either on a screen or as a printout. It displays data flow in both directions of a line as well as timing information. A typical printout is shown in Figure 12.2. This printout has been produced using a software analyser and shows a data signal above a corresponding clock signal.

Protocol analyser

A protocol analyser is an extension of the idea of a data analyser which, as well as displaying data, can also carry out simulations. It normally has two modes of operation, the more basic of which does little more than a data analyser. However, in simulation mode the protocol analyser can simulate the signals and responses expected from a particular piece of equipment or part of the network. In this way it can be used not only to determine the source of failures but also to analyse and hopefully improve network performance. Protocol analysers can be used with most common standard protocols such as X.25 and the IEEE 802 series of protocols.

12.2.3 Trouble ticketing

Trouble tickets have traditionally been a major constituent of network management systems. They are used as a means of logging the problems that arise in a data network. A trouble ticket is an extended log entry containing useful information such as date, time, network device involved and the nature of the problem. It also contains information on any follow-up, such as the remedial action taken and details of any equipment or parts replaced. More recently, paper trouble ticketing has been replaced by trouble ticketing data bases which can play a further role in the management and planning of a network. Thus network managers or engineers may, for example, examine the fault history of a particular item of equipment or the action taken for particular types of faults to see whether the network is functioning efficiently.

12.3 Configuration management

Configuration management involves the long-term planning of a network's topology and inventory of equipment and circuits. Most systems of configuration management contain an inventory database with information on both active and backup devices and connections. As well as aiding the reconfiguration of a network following network failures, configuration management systems allow network managers and engineers to make informed decisions on future network expansion. The inventory may include information on devices which can be configured to perform different functions. For example, the same device might be configured to act either as a router or as an end-system node.

12.4 Accounting management

A network management system needs to keep track of how network resources are used. This is often simply for accounting reasons in networks where individuals or departments are charged for the use of the network resources. Other reasons for accounting management are:

- assisting with the planning of future network development and expansion,
- helping users to make more efficient use of the network,
- the detection of users who may be using the network inappropriately.

12.5 Integrated management systems

Originally, the functions of performance, failure, configuration and accounting management were provided separately for a data communications network. More recently, network management systems integrate these management functions with reporting capabilities in a centralized computer, typically a PC. The computer has access to a variety of local and remote monitoring devices and may itself be connected to a computer network. The monitoring devices used depend on the type of system. For example, a system assembled by a company that supplies routers will use diagnostic monitoring facilities associated with its routers.

12.6 Network management standards

Standards for network management are not as well established as in other areas of data communications. Nevertheless, a number of different committees are working on this issue. The American National Standards Institute (ANSI) committee X3T5.4 has been asked by the ISO to develop a management standard for OSI systems. Within the ISO itself, study group 21 is looking at network management within the OSI model in general. In its capacity as a local area network standardization body, the IEEE 802 committee is working on standards for the management of LANs.

12.6.1 OSI network management

As is the case with most areas of OSI standardization, the series of standards that have been developed by the ISO is both complicated and voluminous. The first of these standards, issued jointly by ISO and the CCITT, was the CCITT X.700 recommendation entitled *OSI Basic Reference Model Part 4: Management Framework* which specified the overall management framework for the OSI model and gave a general introduction to network management concepts. In this context the ISO prefers the term 'systems management' to what is more generally known as network management. A number of standards followed which constitute the CCITT X.700 series of recommendations. Recommendation X.701 provides a general overview of the other standards.

The key elements of the series of documents are recommendations X.710 Common Management Information Services (CMIS) and X.720 Common Management Information Protocols (CMIP). These form the basis for all the OSI network management functions. CMIS is the set of management services provided and CMIP provides a protocol for exchanging information between the points at which management functions are initiated and other points at the same level where the functions are carried out. Five functions known as Specific Management Functional Areas (SMFAs) are specified in the documentation as follows:

(1) *Performance management*   A fairly limited performance management facility is envisaged which will allow for the monitoring and collection of data concerning the current performance of network resources in an open system and also the generation of performance reports. As yet, there is no facility for the prediction of performance patterns. It is possible to make performance predictions using, among other tools, the queuing theory outlined in Appendix A. The lack of this facility is considered by some to be a distinct disadvantage of this area of the OSI management standards.

(2) *Fault management* The OSI fault management facility allows for the detection and identification of abnormal operation in an OSI environment. The specific facilities include the following:

(a) The detection of faults and the passing on of error reports.

(b) The carrying out of diagnostic tests on remote system resources. The ISO uses the term 'managed object' to describe a resource that is managed by its management protocols.

(c) The maintenance of a log of events.

(d) The operation of threshold alarms which are activated if a preset threshold is exceeded or fallen below.

(3) *Configuration management* The OSI configuration management SMFA allows a network manager to observe and modify the configuration of network components in an open system. The following facilities are included:

(a) The collection of data concerning the current configuration of network components.

(b) The alteration of network component configurations.

(c) The initialization and closing down of network components.

(4) *Accounting management* The OSI accounting management SMFA allows network managers to identify the use of resources in an open system and, where costs are incurred in the use of resources, to calculate and allocate such costs. Two main areas of accounting are specified, namely, the transmission system including the communication medium and the end systems. Despite a perceived need for an accounting facility in OSI environments, this aspect of the standards has not, as yet, been fully developed.

(5) *Security management* The term 'security' is frequently associated with sensitive areas such as military systems in which data is often highly confidential or even secret. However, the term has a much wider meaning and a security management facility should ideally include the following functions:

(a) The control of access by users to the various resources of a system.

(b) The protection of both data and operations to ensure that they are carried out correctly.

(c) The authentication of data communications to ensure that data is from the source that it claims to be.

No systems are totally secure. Even a local, isolated system is prone to some insecurity and such problems are multiplied as the size of a network increases. An open system, therefore, is particularly vulnerable in this respect. The ISO has a number of working parties active in the area of security management but progress has so far been slow. The ISO approach is to implement the security management functions within the individual layers of the OSI model. It remains to be seen whether or not this particular SMFA will prove successful.

12.6.2 Network management in the TCP/IP environment

The Transmission Control Protocol/Internet Protocol (TCP/IP) suite of protocols was developed by the US Department of Defense (DoD) as a way of introducing interoperability into the ARPANET wide area network. ARPANET, which was also funded by the DoD, was one of the earliest packet switching networks when it was developed in the early 1970s. The growth of the ARPANET network was rapid and it was clear from an early stage that interoperability would be a problem. This problem has become even greater in recent years as ARPANET has developed into the Internet, which brings together disparate WANs (including ARPANET) and also LANs. The TCP/IP suite of protocols was developed in the late 1970s and, since it was never designed for one particular manufacturer's system, has proved suitable for an internetworking environment in which different networks are linked by gateways and communication takes place between many different manufacturers' computers. The ARPANET network and the TCP/IP evolved for many years without any formal network management capability. It was not until the late 1980s when the Internet started to grow rapidly that attention was given to this matter and a network management protocol called SNMP (Simple Network Management Protocol) was developed for use in a TCP/IP environment. The first SNMP products were produced in 1988 by Cisco Systems among others and its use has spread rapidly since to most major manufacturers. Network management within a TCP/IP environment is based on four key elements:

(1) **Management station**   This is normally a stand-alone device which acts as the interface between a network manager and SNMP. A management station provides typical network management functions such as data analysis, fault recovery and network monitoring facilities. It also has access to a database.

(2) **Management agent**   Other elements within a network such as routers and bridges can be managed from a management station. The active elements within a network that communicate with a management station are known as management agents. Management stations send various commands to management agents which respond appropriately. The agents may also send some important items of information to a management station even though they are not specifically requested.

(3) **Management information base**   At each agent there is a collection of data similar to a database which is known as a management information base. A management station controls an agent by accessing its management information base and retrieving information from it or modifying its contents. The management information bases are distributed throughout a network so a central database known as the **Network Statistics Database** is maintained.

(4) **Network management protocol**   This is the SNMP and it allows the management stations and the agents to communicate with each other.

12.7 Practical network management systems

A number of manufacturers have developed practical network management systems. IBM has two products: NetView is a network management package for use in IBM SNA networks; SystemView is a similar package available to IBM clients who use non-SNA networks. Hewlett Packard has a package called OpenView which has been well received. DEC has a product known as Enterprise Management Architecture, which despite containing an implementation of the OSI standard Common Management Information Protocol (CMIP), is specifically designed for DEC products. All of these products have, to a greater or lesser extent, the same limitation in that they have been developed with a particular manufacturer's system in mind.

As a practical example of a network management system we shall look at a product developed by the SITA group which can claim to be independent of any particular manufacturer's equipment. SITA operates one of the world's largest international communications networks that has developed out of its communications provision for the world's air transport community. SITA's network management package is called SITAVISION, which runs on an IBM-compatible PC with a link to a central server for updating of network information. SITAVISION supports the following types of data connection; X.25, X.28, SDLC (an IBM-based subset of HDLC) and frame relay. The system provides all of the network management functions mentioned in this chapter and serves as a useful illustration of how these functions operate in practice. The main features of the system are now summarized under the headings of these management functions.

12.7.1 Performance management

SITAVISION allows the performance of a network to be monitored by providing the following measurements:

- round-trip delay between two connections
- MTBF and MTTR for all connections
- availability
- logs of node failures

As mentioned in Section 12.1, these measurements help with the planning of future requirements. Figure 12.3 shows a display for round-trip delay between Madrid and the east coast of the United States for a network. The background of the display is an overall view of the network.

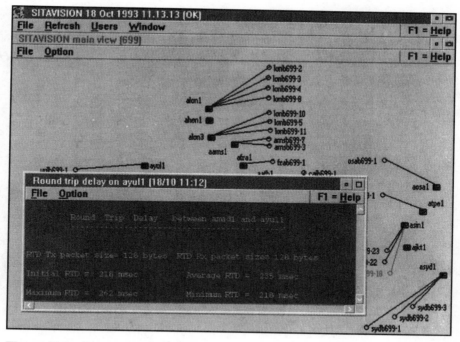

Figure 12.3   Round-trip delay display.

12.7.2 Fault management

In the event of failures, the following three main functions are provided:

- real-time alarm reception which indicates the nature and location of a failure,
- access to any selected link or node so that previous fault conditions can be monitored as well as availabilities and response times,
- real-time monitoring of traffic conditions.

12.7.3 Configuration management

The main view of the network shown in the background of Figure 12.3 represents graphically all nodes and connections. Up to six subviews of these connections can be configured, split by geography, by function or by other criteria. For example, all connections that access one specific computer can be viewed or all connections in a particular region. In addition to graphical views of the network, the following information is available:

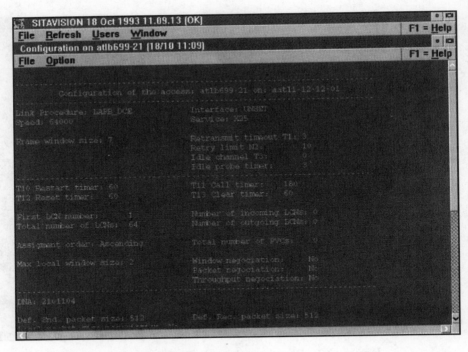

Figure 12.4 Configuration parameters display.

- inventory of connections
- configuration parameters for each connection
- help desk information with local contact details for global support

Figure 12.4 shows a display of configuration parameters for a network connection between nodes atlb and aatl1 in the United States. The display shows a considerable amount of information on the link configuration. The link is being used as part of an X.25 packet switching service and is using the Link Access Protocol-Balanced (LAP-B) protocol which is part of the HDLC protocol. The transmission speed is 64 kbps and the frame window size is 7. Information is given about the logical channel numbers (LCNs) and several other parameters.

12.7.4 Report management

In addition to the three functions just discussed SITAVISION also provides a number of reporting functions which allow statistics about the network to be collated. These statistics can be exported to other applications, such as spreadsheets, for further analysis or presentation. A typical report is shown in Figure 12.5 in the form of a histogram of monthly traffic.

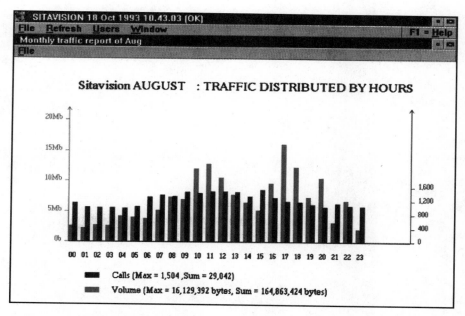

Figure 12.5 Monthly traffic display.

12.8 Future developments

The OSI CMIP network management protocol presents a logical and welcome way forward for the management of open networks. However, its development has so far proved to be slow. SNMP, on the other hand, has stolen a march on CMIP and has become widely used in the same way that TCP/IP has proved so successful. Most major manufacturers, unable to wait for the CMIP to bear fruit, are now supporting SNMP. As long as TCP/IP remains popular, the future of SNMP seems secure and most users will turn to it rather than waiting to see what might develop in the near future. It may be that, with the weight of the ISO behind it, CMIP will win out in the long run but that prospect seems a long way off at the present time.

Applications of queuing theory

A.1 Introduction

An important measure of performance of a data communications network is the average delay while transmitting data from source to destination. Furthermore, delay considerations play an important part in network functions such as flow control and routing. (These are dealt with in some detail in Chapter 7.) Queuing arises naturally in both packet switched and circuit switched networks. Packets arriving at a node in a network will join in a queue whilst waiting to be connected to an outgoing transmission link to the next node along the route. Much of the theory of queuing was developed from the study of telephone traffic at the beginning of the 20th century by A. K. Erlang. The simplest queue, in which data, shown as packets, arrives randomly at an average rate of λ packets per second is shown in Figure A.1. The packets are held in a queue while a server deals with them at a rate of μ per second and then are transmitted.

This type of system is known as a single-server queue, although there is often more than one server in a system. It is important that the arrival rate λ is not allowed to exceed the service rate μ, or the queue will build up (to infinity in theory, but to a maximum size in practice).

A.2 Multiplexing

When a number of streams of data come together over a single link they need to be multiplexed. The most common arrangement in data communications is statistical

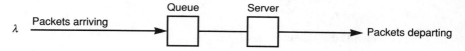

Figure A.1 Single-server queue.

multiplexing in which traffic from a number of data streams is merged into a single queue and then transmitted over the link on a first-come first-served basis. Alternatively, each stream may have its own queue and each queue is served in sequence. If a queue is empty the server moves on directly to the next one. Two types of multiplexing are often used in a circuit switched system. In Time-Division Multiplexing (TDM) the time available on the transmission link is divided into fixed-length time slots which are each allocated to an input stream. In Frequency-Division Multiplexing (FDM) the available frequency bandwidth of the transmission link is divided into fixed frequency bands each of which is allocated to an input stream. The reason for the popularity of statistical multiplexing in data communications is that it produces a smaller average delay. This is because both TDM and FDM allocate time (or frequency) to an empty input stream even if other input streams have data waiting for transmission.

A.3 Little's theorem

This simple theorem provides the basis for much of queuing theory. It arises out of a queuing model in which blocks of data (normally in the form of packets) arrive randomly at a network node. While at a node they are held in a queue awaiting service (retransmission). The time taken to retransmit the packet (equal to the packet length in bits divided by the transmission rate) is often called the 'service time' in this context. The theory applies equally to a circuit switched telephone system in which packets are replaced by calls and the service time is equal to the duration of a call. Little's theorem states that the average number of packets in the system, N, the average delay, T, and the average arrival rate, λ, are related as follows:

$$N = \lambda T \qquad \text{(A.1)}$$

The usefulness of this theorem is that it applies to almost every queuing system. Everyday examples spring to mind. For example, slow-moving traffic (large T) produces crowded streets (large N); a fast-food restaurant (small T) needs fewer tables (small N) than a normal restaurant for the same customer arrival rate (λ). The theorem can also be used to find the average number of packets in a queue rather than the overall system. If we define the following:

W, the average time spent waiting in the queue
N_q, the average number of packets found waiting in the queue by packets on arrival

then Little's theorem leads to:

$$N_q = \lambda W$$

EXAMPLE A.1

A fast-food restaurant is operating with a single person serving customers who arrive at an average rate of 2 per minute and wait to receive their order for an average of 3 minutes. On average, half of the customers eat in the restaurant and the other half take-away. A meal takes an average of 20 minutes to eat. Determine the average number of customers queuing and the average number in the restaurant.

Customers who eat in the restaurant stay on average for 23 minutes, and customers who take-away for 3 minutes. Arrival rate, λ, is 2 per minute.

Average customer time in restaurant, $T = 0.5 \times 23 + 0.5 \times 3 = 13$ minutes.
Average time in queue, $W = 3$ minutes,

From Little's theorem:

Average number of customers queuing, $N_q = \lambda W = 2 \times 3 = 6$
Average number in restaurant, $N = \lambda T = 2 \times 13 = 26$

A.4 Single-server queues

The simplest type of queue is that shown in Figure A.1 in which there is a single server. To take our analysis further it is assumed that the arrival of packets at a network node occurs randomly and independently of each other. This type of arrival process is called a memoryless or Poisson process. The best way of illustrating this is using the probability distribution shown in Figure A.2, where $P(n)$ gives the probability that the number of packets arriving in a particular time interval will equal n.

Suppose now that the packets are dealt with from the queue in the order in which they arrive (first in, first out). The different service times are assumed to be

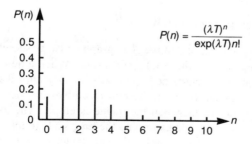

Figure A.2   Poisson probability distribution.

independent of the arrival times and mutually independent. The average service time is denoted by S and the service rate when the server is busy is given by:

$$\mu = \frac{1}{S} \text{ packets per second} \tag{A.3}$$

The average number of packets being retransmitted at any one time equals the arrival rate times the average service time:

$$\lambda S = \frac{\lambda}{\mu}$$

Since one packet is the most that can be transmitted at any one time in single-server systems then this value is also a utilization factor ρ:

$$\rho = \frac{\lambda}{\mu} = S\lambda \tag{A.4}$$

We now introduce the concept of residual service time. By this we mean that if packet i is already being served when packet j arrives, R_j is the time remaining until packet i's service time is complete. If no packet is being served (the queue is empty when packet j arrives) then $R_j = 0$. Thus when a packet arrives at a queue, the average time spent waiting in the queue is equal to the average residual service time, R, plus the product of the average service time and the average number of packets found in the queue:

$$W = R + SN_q \tag{A.5}$$

$$= R + \frac{N_q}{\mu}$$

From Little's theorem:

$$N_q = \lambda W$$

and substituting for N_q gives:

$$W = R + \frac{\lambda}{\mu}W$$

$$= R + \rho W \quad \text{where } \rho \text{ is the utilization factor}$$

Therefore:

$$\text{Average time spent waiting, } W = \frac{R}{1-\rho} \tag{A.6}$$

The simplest way to determine R is graphically, as shown in Figure A.3, by plotting the residual service time $r(t)$ against time. When a new service of duration S begins, the value of $r(t)$ starts at S_1 and falls linearly until it reaches zero after a period of time, S_1. As can be seen, the plot resembles a series of 45° right-angled triangles with gaps occurring when the queue is empty. The average value of $r(t)$ can be found by integrating and extending the time scale to infinity.

This gives the result for average residual service time, R, as:

$$R = \frac{1}{2}\lambda \overline{S^2} \tag{A.7}$$

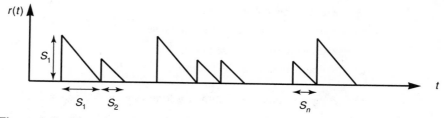

Figure A.3 Plot of residual service time against time.

where $\overline{S^2}$ is the mean square service time. Substituting for R in (A.6) gives:

$$W = \frac{\lambda \overline{S^2}}{2(1-\rho)} \tag{A.8}$$

This is the Pollaczek–Khinchin (P–K) formula which along with Little's theorem provides much of the theoretical basis for queuing theory.

The total time spent both waiting in the queue and being serviced (transmitted) can now be written as:

$$T = S + \frac{\lambda \overline{S^2}}{2(1-\rho)} \tag{A.9}$$

The average number of items in the queue, N_q, and the average number in the system, N, can be obtained by applying Little's theorem to (A.8) and (A.9):

$$N_q = \frac{\lambda^2 \overline{S^2}}{2(1-\rho)} \tag{A.10}$$

$$N = S\lambda + \frac{\lambda^2 \overline{S^2}}{2(1-\rho)}$$

But from (A.4), $S\lambda = \rho$, so that:

$$N = \rho + \frac{\lambda^2 \overline{S^2}}{2(1-\rho)} \tag{A.11}$$

This P–K derivation assumes that packets are served in the order of arrival (FIFO). However, the formula holds even if packets are served in a different order as long as the order of service is independent of the service times.

EXAMPLE A.2

Packets arrive at a single-server node for transmission at random with an average arrival rate of 15 per second. 80% of packets contain 1024 bits and 20% contain 64 bits. If the transmission rate is 19.2 kbps and the system is error free, determine:

(a)   average transmission time

(b)   average time spent waiting in queue

(c)   average number in the queue

(d)   average number in the system

(a)   Average transmission time for long packets $= \dfrac{1024}{19.2k} = 53.3\,\text{ms}$

Average transmission time for short packets $= \dfrac{64}{19.2k} = 3.3\,\text{ms}$

Average transmission time $= 0.8 \times 53.3 + 0.2 \times 3.3$
$$S = 43.3\,\text{ms}$$

(b)   Mean square service time $= 0.8 \times (53.3)^2 + 0.2(3.3)^2$
$$= 2275.5 + 2.2$$
$$= 2277.7\,\mu\text{s}^2$$

(Note that in calculating the mean square service time, the short packets could have been ignored without much affecting the result. This is not always the case, particularly when errors are taken into account.)

Utilization factor, $\rho = \lambda S$
$$= 15 \times 43.3 \times 10^{-3}$$
$$= 0.65$$

The average time spent waiting in the queue is obtained directly from P–K:

$$W = \frac{\lambda \overline{S^2}}{2(1 - \rho)} = \frac{15 \times 2278 \times 10^{-6}}{2 \times 0.35} = 48.8\,\text{ms}$$

(c)   The average number in the queue is obtained from Little's theorem:

$N_q = \lambda W$
$\quad = 15 \times 48.8 \times 10^{-3}$
$\quad = 0.732$ packets

(d)   Average number of packets in the system:

$N = S\lambda + N_q$
$\quad = 43.3 \times 15 + 0.732$
$\quad = 0.649 + 0.732$
$\quad = 1.3815$

A.5 Further queue types

The Pollaczek–Khinchin formula can be used to analyse most types of queue. Each type of queue is specified by three letters and numbers. The type considered up to now is M/G/1 which has the following meaning:

(1)   The first letter gives the nature of the arrival process: M for memoryless (that is, a Poisson probability distribution); G for general (unspecified); D for deterministic (at predetermined time intervals).

(2)   The second letter stands for the nature of the probability distribution of the service (that is, transmission times).

(3)   The final number is the number of servers.

A.5.1   The M/M/1 queuing system

This is similar to the M/G/1 system discussed above except that the service process has the same memoryless (Poisson) process as the arrival process. This has the effect of giving the service times an exponential probability distribution, as shown in Figure A.4.

This is often considered the worst-case situation since it involves a wide range of service times. The mean square value of the service time can be determined from the exponential distribution. This gives the following result:

$$\text{Mean square service time, } \overline{S^2} = \frac{2}{\mu^2} \qquad\qquad\text{(A.12)}$$

The P–K formula and related equations then become:

$$\text{Pollaczek–Khinchin, } W = \frac{\lambda 2}{\mu^2 2(1-\rho)}$$

$$= \frac{\rho}{\mu(1-\rho)}$$

$$\text{Average time spent in the system, } T = S = \frac{\rho}{\mu(1-\rho)} = \frac{1}{\mu} + \frac{\rho}{\mu(1-\rho)}$$

$$= \frac{1-\rho+\rho}{\mu(1-\rho)} = \frac{1}{\mu(1-\rho)}$$

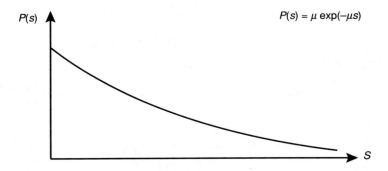

Figure A.4   Exponential probability distribution.

Average number waiting in queue, $N_q = \lambda W = \dfrac{\lambda \rho}{\mu(1 - \rho)}$

$$= \frac{\rho^2}{1 - \rho} \qquad \text{since } \frac{\lambda}{\mu} = \rho$$

Average number in system, $N = \lambda T = \dfrac{\lambda}{\mu(1 - \rho)}$

$$= \frac{\rho}{1 - \rho}$$

EXAMPLE A.3

A node in a half-duplex multipoint system has the following characteristics:

Single server
Arrival rate, $\lambda = 750$ messages per hour
Average transmission time, $S = 1.5$ seconds

Determine the worst-case average time spent waiting in queue and the corresponding average number of messages waiting in queue.

Since a worst-case value is required, we assume that an M/M/1 queue is used.

Utilization factor, $\rho = \lambda S$
$$= \frac{750 \times 1.5}{3600}$$
$$= 0.3125$$

Service rate, $\mu = \dfrac{\lambda}{\rho} = \dfrac{750}{3600 \times 0.3125}$
$$= 0.667 \text{ per second}$$

Mean square service time, $\overline{S^2} = \dfrac{2}{\mu^2}$ \qquad (Equation A.12)
$$= 2/0.667^2 = 4.5 \, \text{s}^2$$

From the Pollaczek–Khinchin formula:

Average time waiting in queue, $W = \dfrac{\lambda \overline{S^2}}{2(1 - \rho)} = \dfrac{750 \times 4.5}{3600 \times 2 \times (1 - 0.3125)}$
$$= 0.68 \, \text{s}$$

Average number in queue, $N_q = \lambda W = \dfrac{750}{3600} \times 0.68$
$$= 0.142$$

It can be seen that messages spend an average of 0.68 seconds waiting in the queue, which is quite a long time in a data communications system, yet the utilization factor is only 0.3125 which is fairly low. Why is this? The reason is that the average transmission time of 1.5 seconds is slow. To improve the efficiency and throughput of this system it is necessary to increase the transmission rate of the half-duplex link.

A.5.2 The M/D/1 queuing system

This type of system has predetermined equal service times for all packets and thus gives a best-case situation since there is no deviation from an average value of service time. The mean square value of service time is given by:

$$\overline{S^2} = S^2$$

$$= \frac{1}{\mu^2} \qquad \text{from (A.3)}$$

Note that this is half the value of the M/M/1 case.

This gives a result from the Pollaczek–Khinchin formula of:

Average time in queue, $W = \dfrac{\lambda}{\mu^2 \, 2(1-\rho)}$

$$= \frac{\rho}{2\mu(1-\rho)}$$

Average time in the system, $T = S + \dfrac{\rho}{2\mu(1-\rho)}$

$$= \frac{1}{\mu} + \frac{\rho}{2\mu(1-\rho)}$$

$$= \frac{2(1-\rho) + \rho}{2\mu(1-\rho)}$$

$$= \frac{2-\rho}{2\mu(1-\rho)}$$

Average number in queue, $N_q = \lambda W = \dfrac{\lambda\rho}{2\mu(1-\rho)}$

$$= \frac{\rho^2}{2(1-\rho)}$$

Average number in system, $N = \lambda T = \dfrac{\lambda(2-\rho)}{2\mu(1-\rho)}$

$$= \frac{\rho(2-\rho)}{2(1-\rho)}$$

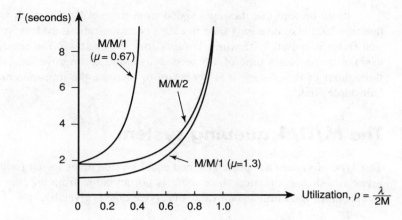

Figure A.5 Plot of time delay against utilization.

A.5.3 Multiple-server queues

In practice some nodes in a system may have more than a single outgoing transmission link. For example, consider the case of two outgoing transmission links connecting a packet switching node to a neighbouring node. Packets will use either link randomly. Such a system will use an M/M/2 queuing system at the first node. This will produce double the service rate as an M/M/1 queue and will be able to handle twice the arrival rate. Adding a second server thus improves both the time delays and throughput performance. This is illustrated in Figure A.5, showing time delay, T, plotted against utilization for the M/M/1 queue of Example A.3 along with the corresponding M/M/2 queue and a M/M/1 queue with doubled service rate.

Note that the M/M/2 queue gives lower delay times. It is clear, however, from the corresponding curve for an M/M/1 queue with a doubled service capacity that this queue outperforms the M/M/2 queue at low utilizations since, when traffic is low, only one of the two links in the M/M/2 system is being used. As a general rule it is better to double the transmission capacity of a link rather than to add a second link.

A.6 Effects of errors on delays

If errors occur in a system and Automatic Repeat on reQuest (ARQ) is used for error correction to retransmit erroneous packets, then, the average transmission time will increase. This will, in turn, result in increased delays and queue lengths. In determining an average transmission time, both the error rate and the type of ARQ strategy need to be taken into account. The greater the error rate, the more packets will need to be retransmitted, and the greater will be transmission and queuing times and queue lengths.

A.7 Networks of transmission links

In a data communications network there are large numbers of transmission queues which interact with each other, in that traffic leaving a queue at one node may enter one or more other queues, possibly merging with traffic leaving further queues on the way. This tends to complicate the arrival process at the subsequent queues. Consequently it is not possible to use the (relatively) simple type of analysis of queuing systems explored here. However, computer simulation packages are available, such as CACI's Comnet III, which use queuing models as part of their network simulation. These packages can be run on PCs, but become slow and expensive in computer time if the network modelled is large.

A.7 Networks of transmission links

Glossary

Advanced Research Projects Agency (ARPA) *see* ARPANET

American Standards Committee for Information Interchange (ASCII) In normal usage this refers to the character code defined by this committee for the interchange of information between two communicating devices. The ASCII character set is in widespread use for the transfer of information between a computer and a peripheral device such as a visual display unit or a printer.

Amplitude modulation (AM) A modulation technique to allow data to be transmitted across an analogue network, such as a switched telephone network. The amplitude of a single (carrier) frequency is varied (modulated) between two levels – one for binary 0 and the other for binary 1.

ANSI American National Standards Institute.

Application layer This corresponds to layer 7 of the ISO reference model for open systems interconnection. It comprises a number of application-oriented protocols and forms the user interface to the various distributed information processing services supported.

ARPA *see* ARPANET

ARPANET The wide area network funded by the former Advanced Research Projects Agency (ARPA) and now known as the Defense Advanced Projects Research Agency (DARPA).

ASCII American standard code for information interchange.

Association control service element (ACSE) A protocol entity forming part of the application layer. It provides the generalized (common) function of establishing and clearing a logical association (connection) between two application entities.

Asynchronous transfer mode (ATM) The proposed mode of operation of the emerging broadband integrated services digital network. All information to be transmitted – voice, data, image, video – is first fragmented into small, fixed-sized frames known as cells. These are switched and routed using packet switching principles – also known as cell or fast-packet switching. The first networks that are based on this mode of operation are ATM LANs.

Asynchronous transmission Strictly, this implies that the receiver clock is not synchronized to the transmitted clock when data is being transmitted between two devices connected by a transmission line. More generally, it indicates that data is being transmitted as individual characters. Each character is preceded by a start signal and terminated by one or more stop signals, which are used by the receiver for synchronization purposes.

Automatic repeat request (ARQ)   A technique used for error control over a transmission line. If errors in a transmitted message are detected by the receiving device, it requests the sending device to retransmit the message together with any other messages that might have been affected.

Bandwidth   The difference between the highest and the lowest sinusoidal frequency signals that can be transmitted across a transmission line or through a network. It is measured in hertz (Hz) and also defines the maximum information-carrying capacity of the line or network.

Baseband   A particular operating mode of a transmission line: each binary digit (bit) in a message is converted into one of two voltage (sometimes current) levels – one for binary 1 and the other for binary 0. The voltages are then applied directly to the line. The line signal varies with time between these two voltage levels as the data is transmitted.

Baud   The number of line signal variations per second. It also indicates the rate at which data is transmitted on a line, although this is strictly correct only when each bit is represented by a single signal level on the transmission line. Hence, the bit rate and the line signal rate are both the same.

BER   *see* Bit error rate

Bit error rate (BER)   The probability that a single bit, when transmitted over a link, will be received in error.

Bit stuffing (zero bit insertion)   A technique used to allow pure binary data to be transmitted on a synchronous transmission line. Each message block (frame) is encapsulated between two flags, which are special bit sequences. If the message data contains a possibly similar sequence, then an additional (zero) bit is inserted into the data stream by the sender, and is subsequently removed by the receiving device. The transmission method is said to be data transparent.

Bridge   A device used to link two homogeneous local area subnetworks, that is, two subnetworks utilizing the same physical and medium access control method.

Broadband   A particular mode of operation of a coaxial cable. A single coaxial cable can be used to transmit a number of separate data streams simultaneously by assigning each stream a portion of the total available bandwidth. Data is transmitted by modulating a single frequency signal from the selected frequency band and is received by demodulating the received signal.

Broadcast   A means of transmitting a message to all devices connected to a network. Normally, a special address, the broadcast address, is reserved to enable all the devices to determine that the message is a broadcast message.

Bus   A network topology in widespread use for the interconnection of communities of digital devices distributed over a localized area. The transmission medium is normally a single coaxial cable to which all the devices are attached. Each transmission thus propagates the length of the medium and is received by all other devices connected to the medium.

CCITT   International Telegraph and Telephone Consultative Committee (now ITU-T).

CCR   Commitment, concurrency and recovery.

Circuit switching   The mode of operation of a telephone network and also some of the newer digital data networks. A communication path is first established through the network between the source (calling) and destination (called) terminals, and this is used exclusively for the duration of the call or transaction. Both terminals must operate at the same information transfer rate.

CMIS *see* Common management information system

Coaxial cable A type of transmission medium consisting of a centre conductor and a concentric outer conductor. It is used when high data transfer rates (greater than 1 Mbps) are required.

Commitment, concurrency and recovery (CCR) A protocol entity forming part of the application layer. It allows two or more application processes to perform mutually exclusive operations on shared data. It also provides control to ensure that the operations are performed either completely or not at all. It uses the concepts of an atomic action and a two-phase commit protocol.

Common management information protocol (CMIP) The ISO application layer protocol used to retrieve and send management-related information across an OSI network.

Common management information system (CMIS) The set of management services provided in the ITU-T X.710 recommendation which forms the basis for the OSI network management functions.

Community antenna television (CATV) A facility used in the context of local area data networks, since the principles and network components used in CATV networks can also be used to produce a flexible underlying data transmission facility over a local area. CATV networks operate using the broadband mode of working.

Concentrator A process whereby a number of data sources are combined in order to share a single channel to improve channel utilization.

CRC *see* Cyclic redundancy check

Crosstalk An unwanted signal that is picked up in a conductor as a result of some external electrical activity.

CSMA/CD Carrier sense, multiple access with collision detection. A method used to control access to a shared transmission medium, such as a coaxial cable bus to which a number of stations are connected. A station that wishes to transmit a message first senses (listens to) the medium and transmits the message only if the medium is quiet – no carrier present. Then, as the message is being transmitted, the station monitors the actual signal on the transmission medium. If this is different from the signal being transmitted, a collision is said to have occurred and been detected. The station then ceases transmission and tries again later.

Cyclic code A type of block error detection and correction code used extensively in data communication networks. The code words of a cyclic code are distinguished in that if their bits are shifted, either right or left, they produce another code word.

Cyclic redundancy check (CRC) A technique used for the detection of errors in transmitted data which uses a cyclic code. A numeric value is computed from the bits in the data to be transmitted which is placed in the trailer of a block of data. A receiver is able to detect the presence, or otherwise, of errors by repeating the check.

DARPA *see* ARPANET

Data circuit terminating equipment (DCE)   The equipment provided by the network authority (provider) for the attachment of user devices to the network. It takes on different forms for different network types.

Data link layer   This corresponds to layer 2 of the ISO reference model for open systems interconnection. It is concerned with the reliable transfer of data (no residual transmission errors) across the data link being used.

Data terminal equipment (DTE)   A generic name for any user device connected to a data network. It includes such devices as visual display units, computers and office workstations.

Datagram   A type of service offered on a packet switched data network (*see also* Virtual circuit). A datagram is a self-contained packet of information that is sent through the network with minimum protocol overheads.

Defense Advanced Research Projects Agency (DARPA)   *see* ARPANET

Delay distortion   Distortion of a signal caused by the frequency components making up the signal having different propagation velocities across a transmission medium.

Demultiplexer   Performs the opposite process to that of a multiplexer.

Digital signal processing (DSP)   A digital technique, often by means of a computer, to process signals which are in sampled data representation. Analogue signals are often converted into such a form by means of an analogue-to-digital convertor (ADC) to enable processing by means of DSP.

Distributed queue, dual bus (DQDB)   An optical fibre-based network that can be used as a high-speed LAN or MAN that is compatible with the evolving broadband ISDN. It operates in a broadcast mode by using two buses, each of which transmits small, fixed-sized frames – known as cells – in opposite directions. Each bus can operate at hundreds of megabits per second.

DNA   Digital network architecture.

DSP   *see* Digital signal processing

EIA   Electrical Industries Association.

EIA-232D   Standard laid down by the American EIA for interfacing a digital device to a PTT-supplied modem. Also used as an interface standard for connecting a peripheral device, such as a visual display unit or a printer, to a computer.

Entropy   Average information content of a code word.

Equalizer   A circuit that compensates for imperfections in the amplitude and phase characteristics of a channel.

Ethernet   The name of the LAN invented at the Xerox Corporation Palo Alto Research Center. It operates using the CSMA/CD medium access control method. The early specification was refined by a joint team from Digital Equipment Corporation, Intel Corporation and Xerox Corporation and this in turn has now been superseded by the IEEE 802.3 – ISO 8802.3 – international standard.

ETSI   European Telecommunication Standards Institute.

Extended binary coded decimal interchange code (EBCDIC)   The character set used on all IBM computers.

Fibre distributed data interface (FDDI)   An optical fibre-based ring network that can be used as a high-speed LAN or MAN. It provides a user bit rate of 100 Mbps and uses a control token medium access control method.

Fibre optic   *see* Optical fibre

File transfer access and management (FTAM)   A protocol entity forming part of the application layer. It enables user application processes to manage and access a (distributed) file system.

Flow control   A technique to control the rate of flow of frames or messages between two communicating entities.

Frame   The unit of information transferred across a data link. Typically, there are control frames for link management and information frames for the transfer of message data.

Frame check sequence (FCS)   A general term given to the additional bits appended to a transmitted frame or message by the source to enable the receiver to detect possible transmission errors.

Frame relay   A recently introduced alternative packet switched data service operating at higher speed than X.25 networks.

Frequency-shift keying   A modulation technique to convert binary data into an analogue form comprising two sinusoidal frequencies. It is widely used in modems to allow data to be transmitted across a (analogue) switched telephone network.

FTAM   File transfer access and management.

Full-duplex   A type of information exchange strategy between two communicating devices whereby information (data) can be exchanged in both directions simultaneously. It is also known as two-way simultaneous.

Gateway   A device that routes datagrams (packets) between one network and another. Typically, the two networks operate with different protocols, and so the gateway also performs the necessary protocol conversion functions.

Half-duplex   A type of information exchange strategy between two communicating devices whereby information (data) can be exchanged in both directions alternately. It is also known as two-way alternate.

High-level data link control (HDLC)   An internationally agreed standard protocol defined to control the exchange of data across either a point-to-point data link or a multidrop data link.

Host   Normally a computer belonging to a user that contains (hosts) the communication hardware and software necessary to connect the computer to a data communication network.

IEEE   Institute of Electrical and Electronic Engineers.

Integrated services digital network (ISDN)   The new generation of worldwide telecommunications network that utilizes digital techniques for both transmission and switching. It supports both voice and data communications.

International alphabet number 5 (IA5)   The standard character code defined by ITU-T and recommended by ISO. It is almost identical to the ASCII code.

Internet   The abbreviated name given to a collection of interconnected networks. Also, the name of the US government funded internetwork based on the TCP/IP suite.

Internet protocol (IP)   The TCP/IP that provides connectionless network service between multiple packet switched networks interconnected by gateways.

Intersymbol interference (ISI)   Caused by delay distortion introduced in transmission which results in a dispersion (in time) of a received symbol such that it overlaps into the time periods of adjacent symbols.

ISI   *see* Intersymbol interference

ISO   International Standards Organization.

ITU   International Telecommunications Union.

JTAM   Job transfer, access and management.

Link management   A function of the data link layer of the OSI reference model which is concerned with setting up and disconnection of a link.

Local area network (LAN)   A data communication network used to interconnect a community of digital devices distributed over a localized area of up to, say, 10km^2. The devices may be office workstations, mini- and microcomputers, intelligent instrumentation equipment, and so on.

Logical link control (LLC)   A protocol forming part of the data link layer in LANs. It is concerned with the reliable transfer of data across the data link between two communicating systems.

Management information base (MIB)   The name of the database used to hold the management information relating to a network or internetwork.

Mark   A term traditionally used in telegraph systems to indicate a logic 1 state of a bit.

Medium access control (MAC)   Many LANs use a single common transmission medium – for example, a bus or ring – to which all the interconnected devices are attached. A procedure must be followed by each device, therefore, to ensure that transmissions occur in an orderly and fair way. In general, this is known as the medium access control procedure. Two examples are CSMA/CD and (control) token.

Metropolitan area network (MAN)   A network that links a set of LANs that are physically distributed around a town or city.

Modem   The device that converts a binary (digital) data stream into an analogue (continuously varying) form, prior to transmission of the data across an analogue network (MODulator), and reconverts the received signal back into its binary form (DEModulator). Since each access port to the network normally requires a full-duplex (two-way simultaneous) capability, the device must perform both the MODulation and the DEModulation functions; hence the single name MODEM is used. As an example, a modem is normally required to transmit data across a telephone network.

Multidrop   A type of network configuration that supports more than two stations on the same transmission medium.

Multiplexer   A device to enable a number of lower bit rate devices, normally situated in the same location, to share a single higher bit rate transmission line. The data-carrying capacity of the latter must be in excess of the combined bit rates of the low bit rate devices.

Network layer   This corresponds to layer 3 of the ISO reference model for open systems interconnection. It is concerned with the establishment and clearing of logical or physical connections across the network being used.

Network management   A generic term embracing all the functions and entities involved in the management of a network. This includes configuration management, fault handling and the gathering of statistics relating to usage of the network.

Noise   The extraneous electrical signals that may be generated or picked up in a transmission line. Typically, it may be caused by neighbouring electrical apparatus. If the noise signal is large compared with the data-carrying signal, the latter may be corrupted and result in transmission errors.

NRZ/NRZI   Two similar (and related) schemes for encoding a binary data stream. The first has the property that a signal transition occurs whenever a binary 1 is present in the data stream and the second whenever a binary 0 is present. The latter is utilized with certain clocking (timing) schemes.

Open system   A vendor-independent set of interconnected computers that all utilize the same standard communications protocol stack based on either the ISO/OSI protocols or TCP/IP.

Open systems interconnection (OSI)   The protocol suite that is based on ISO protocols to create an open systems interconnection environment.

Optical fibre   A type of transmission medium over which data is transmitted in the form of light waves or pulses. It is characterized by its potentially high bandwidth, and hence data-carrying capacity, and its high immunity to interference from other electrical sources.

PABX   Private automatic branch exchange.

Packet assembler/disassember (PAD)   A device used with an X.25 packet switching network to allow character-mode terminals to communicate with a packet-mode device, such as a computer.

Packet switching   A mode of operation of a data communication network. Each message to be transmitted through the network is first divided into a number of smaller, self-contained message units known as packets. Each packet contains addressing information. As each packet is received at an intermediate node (exchange) within the network, it is first stored and, depending on the addressing information contained within it, forwarded along an appropriate link to the next node and so on. Packets belonging to the same message are reassembled at the destination. This mode of operation ensures that long messages do not degrade the response time of the network. Also, the source and destination devices may operate at different data rates.

PAM   Pulse amplitude modulation.

Parity   A mechanism used for the detection of transmission errors when single characters are being transmitted. A single binary bit, known as the parity bit, the value (1 or 0) of which is determined by the total number of binary 1s in the character, is transmitted with the character so that the receiver can determine the presence of single-bit errors by comparing the received parity bit with the (recomputed) value it should be.

PCM   Pulse code modulation.

Phase-shift keying (PSK)　A modulation technique to convert binary data into an analogue form comprising a single sinusoidal frequency signal with a phase that varies according to the data being transmitted.

Physical layer　This corresponds to layer 1 of the ISO reference model for open systems interconnection. It is concerned with the electrical and mechanical specification of the physical network termination equipment.

Piggybacking　A technique to return acknowledgement information across a full-duplex (two-way simultaneous) data link without the use of special (acknowledgement) messages. The acknowledgement information relating to the flow of messages in one direction is embedded (piggybacked) into a normal data-carrying message flowing in the reverse direction.

Pixel　The smallest picture element, or cell, which may be physically resolved on a CRT screen.

Postal, Telegraph and Telephone (PTT)　The administrative authority that controls all the postal and public telecommunications networks and services in a country.

POTS　Plain old telephone system.

Presentation layer　This corresponds to layer 6 of the ISO reference model for open systems interconnection. It is concerned with the negotiation of a suitable transfer (concrete) syntax for use during an application session and, if this is different from the local syntax, for the translation to and from this syntax.

Primitive　A type of PDU, such as a request or response.

Protocol　A set of rules formulated to control the exchange of data between two communicating parties.

Protocol data unit (PDU)　The message units exchanged between two protocol entities.

PSTN　Public switched telephone network.

Public data network (PDN)　A packet switched communication network set up and controlled by a public telecommunications authority for the exchange of data.

Remote operations service element (ROSE)　A protocol entity forming part of the application layer. It provides a general facility for initiating and controlling operations remotely.

Ring　A network topology in widespread use for the interconnection of communities of digital devices distributed over a localized area, such as a factory or a block of offices. Each device is connected to its nearest neighbour until all the devices are connected in the form of a closed loop or ring. Data is transmitted in one direction only and, as each message circulates around the ring, it is read by each device connected in the ring. After circulating around the ring, the source device removes the message from the ring.

Router　A device used to interconnect two or more LANs together, each of which operates with a different medium access control method – also known as a gateway or intermediate system.

Server　A facility found in many local area networks where file access, printing or communication functions are provided to other stations.

Service access point (SAP)　The subaddress used to identify uniquely a particular link between two protocol layers in a specific system.

Session layer This corresponds to layer 5 of the ISO reference model for open systems interconnection. It is concerned with the establishment of a logical connection between two application entities and with controlling the dialogue (message exchange) between them.

Simple network management protocol (SNMP) The application protocol in a TCP/IP suite used to send and retrieve management-related information across a TCP/IP network.

Simplex A type of information exchange strategy between two communicating devices whereby information (data) can be passed only in one direction.

Slotted ring A type of local area (data) network. All the devices are connected in the form of a (physical) ring and an additional device known as a monitor is used to ensure that the ring contains a fixed number of message slots (binary digits) that circulate around the ring in one direction only. A device sends a message by placing it in an empty slot as it passes. This is read by all other devices on the ring and subsequently removed by the originating device.

SNA Systems network architecture.

Space A term traditionally used in telegraph systems to indicate a logic 0 state of a bit.

Star A type of network topology in which there is a central node that performs all switching (and hence routing) functions.

Statistical multiplexer (stat mux) A device used to enable a number of lower bit rate devices, normally situated in the same location, to share a single, higher bit rate transmission line. The devices usually have human operators, and hence data is transmitted on the shared line on a statistical basis rather than, as is the case with a basic multiplexer, on a preallocated basis. It endeavours to exploit the fact that each device operates at a much lower mean rate than its maximum rate.

Synchronous transmission A technique to transmit data between two devices connected by a transmission line. The data is normally transmitted in the form of blocks, each comprising a string of binary digits. With synchronous transmission, the transmitter and receiver clocks are in synchronism; a number of techniques are used to ensure this.

Syndrome The result of a computation undertaken to detect whether or not errors have occurred in a data transmission which uses an error detection or correction code.

TCP/IP The complete suite of protocols, including IP, TCP and the associated application protocols.

Teletex An international telecommunications service that provides the means for messages, comprising text and selected graphical characters, to be prepared, sent and received.

Time-division multiplexing (TDM) A technique to share the bandwidth (channel capacity) of a shared transmission facility to allow a number of communications to be in progress either concurrently or one at a time.

Token bus A type of local area (data) network. Access to the shared transmission medium, which is implemented in the form of a bus to which all the communicating devices are connected, is controlled by a single control (permission) token. Only the current owner of the token is allowed to transmit a message on the medium. All devices wishing to transmit messages are connected in the form of a logical ring. After a device receives the token and transmits any waiting messages, it passes the token to the next device on the ring.

Token ring   A type of local area (data) network. All the devices are connected in the form of a (physical) ring and messages are transmitted by allowing them to circulate around the ring. A device can transmit a message on the ring only when it is in possession of a control (permission) token. A single token is passed from one device to another around the ring.

Transmission control protocol (TCP)   The protocol in the TCP/IP suite that provides a reliable full-duplex message transfer service to application protocols.

Transmission medium   The communication path linking two communicating devices. Examples are metallic or optical cable.

Transport layer   This corresponds to layer 4 of the ISO reference model for open systems interconnection. It is concerned with providing a network-independent, reliable message interchange service to the application-oriented layers (layers 5 through 7).

Unshielded twisted pair (UTP)   A type of transmission medium consisting of two insulated wires, normally twisted together in order to reduce interference due to induction from electromagnetic fields in close proximity, but without an external shield.

Videotex   A telecommunications service that allows users to deposit and access information to and from a central database facility. Access is through a special terminal comprising a TV set equipped with a special decoder.

Virtual circuit   A call request packet sets up a suitable route through the network. All subsequent packets from the same source then follow the same route through the network, which is now called a virtual circuit. These packets do not, however, have exclusive use of the circuit. Packets from a number of different sources may share the use of each link.

Wide area network (WAN)   Any form of network – private or public – that covers a wide geographical area.

Wireless LAN   A LAN that uses either radio or infrared as the transmission medium. Requires different MAC methods from those used with wired LANs.

Bibliography

Bertsekas D. and Gallager R. (1987). *Data Networks*. Prentice Hall

Black U. D. (1989). *Data Networks: Concepts, Theory and Practice*. Prentice Hall

Brewster. R.L. (1989). *Data Communications and Networks 2*. Peter Peregrinus

Clark G.C. and Cain J.B. (1981). *Error-Correction Coding for Digital Communications*. Plenum Press

Conard (1980). Character oriented data link control procedures. *IEEE Trans.*, April

Cuthbert L.G. and Sapanel J.C. (1993). *ATM: The Broadband Telecommunications Solution*. IEE

Datapro Research Corp. (1987). *Network Management Systems*.

De Pryckner M. (1991). *Asynchronous Transfer Mode: Solution for Broadband ISDN*. Ellis Horwood

Dettmer R. (1992). Frame relay the network express. *IEE Review*, (November/December), 381-5

Freer J. (1988). *Computer Communications and Networks*. Pitman

Friend G. E. *et al.* (1984). *Understanding Data Communications*. Howard W. Sams & Co.

Gibson R. (1989). RS 232C for instrument control. *Electronic and Wireless World*, (September) 890–1

Gray J. (1972). Line control procedures. *Proc. IEEE* (November), 1301–12

Griffiths J.M. ed. (1992). *ISDN Explained* 2nd edn. John Wiley

Halsall F. (1996) *Data Communications, Computer Networks and Open Systems* 4th edn. Addison-Wesley

Hamming R.W. (1950). Error detecting and error correcting codes. *Bell Syst. Tech. J.* **29**, 147–60

Handel R., Huber M.N. and Schroeder S. (1994). *ATM Networks: Concepts, Protocols, Applications*. Addison-Wesley

Held G. (1986). *Data Communications Networking Devices*. John Wiley

Helgert H.J. (1991). *Integrated Services Digital Network*. Addison-Wesley

Housley T. (1987). *Data Communications and Teleprocessing Systems*. Prentice Hall

IEEE (1977). Flow control: a comparative survey. *IEEE Trans.*, (COM-25), 48–60

Jain R. (1993). *FDDI Handbook High-Speed Networking Using Fiber and Other Media*. Addison-Wesley

Kauffels F.J. (1992). *Network Management*. Addison-Wesley

Lane J.E. (1987). *Packet Switch Stream PSS*. NCC Publications

Lechleider J.W. (1989). Line codes for digital subscriber lines. *IEEE Communication*, (Sep), 23–32.

Marshall. G. (1983). *Principles of Digital Communication Systems*. McGraw-Hill

Martin J., Chapman K., Cooperative Processes Inc., *et al. Local Area Networks Architectures and Implementations*. Prentice Hall

Mazda F. ed. (1993). *Telecommunications Engineers' Reference Book*. Butterworth-Heinemann

Meggitt J.E. (1961). Error correcting codes and their implementation for data transmission systems. *IRE Trans. Inf. Theory*, (IT-7), 234–44

Open University. (1990). *The Open University Digital Telecommunications Course*, T322

Peterson W. W. (1961). *Error Correcting Codes*. Cambridge: MIT Press

Pieder K., Cronin W.J. Jnr. and Michael W.A. (1992). *FDDI for Local Area Networks*. Prentice Hall

Piggot C. W. (1991). What's confusing about RS232C IEEIE. *Electrotechnology*, (February/March) 18–21

Pretzel O. (1992). *Error Correcting Codes and Finite Fields*. Clarendon Press

Read R. J., Bishop P. G. and Jones E. V. (1994). Modems for the general switched telephone network: development design and future. *Electronics & Communications*, **6** (1), 40–8

Redmill F. and Valdar A. (1994). *SPC Digital Telephone Exchanges*. IEE

Scitor Ltd. (1993). *SITAVISION – Effective Network Management*.

Shanmugam K.S. (1979). *Digital and Analog Communication Systems*. John Wiley

Stallings W. (1993). *Networking Standards: A Guide to OSI, ISDN, LAN and MAN Standards*. Addison-Wesley

Stallings W. (1991). *Data and Computer Communications* 3rd edn. Macmillan

Stallings W. (1993). *SNMP, SMP, and CMIP*. Addison-Wesley

Stallings W. (1994). *Data and Computer Communications* 4th edn. Macmillan

Stone H. S. (1982). *Microcomputer Interfacing*. Addison-Wesley

Waters A. G. (1991). *Computer Communication Networks*. McGraw-Hill

Young P.H. (1994). *Electronic Communications Techniques* 3rd edn. Macmillan

Index